研究生教学用书

武汉大学研究生教学用书出版基金资助

表面活性剂化学
SURFACTANT CHEMISTRY

中英文对照

Julian Eastoe 著

武汉大学化学与分子科学学院胶体与界面科学实验室 译

武汉大学出版社

图书在版编目(CIP)数据

表面活性剂化学/(英)Julian Eastoe 著;武汉大学化学与分子科学学院胶体与界面科学实验室译. —武汉:武汉大学出版社,2005.11
研究生教学用书
ISBN 7-307-04552-4

Ⅰ.表… Ⅱ.①J… ②武… Ⅲ.表面活性剂—双语教学—高等学校—教材 Ⅳ.TQ423

中国版本图书馆 CIP 数据核字(2005)第 033881 号

责任编辑:谢文涛 责任校对:王 建 版式设计:支 笛

出版发行:武汉大学出版社 (430072 武昌 珞珈山)
(电子邮件:wdp4@whu.edu.cn 网址:www.wdp.com.cn)
印刷:武汉大学出版社印刷总厂
开本:787×980 1/16 印张:12.125 字数:217 千字
版次:2005 年 11 月第 1 版 2005 年 11 月第 1 次印刷
ISBN 7-307-04552-4/O·319 定价:20.00 元

版权所有,不得翻印;凡购我社的图书,如有缺页、倒页、脱页等质量问题,请与当地图书销售部门联系调换。

前　言

　　2002年10月,我应邀在太原中国日用化学研究院作关于表面活性剂的系列讲座。中国日用化学研究院是中国在胶体科学基础、实践方面最重要的研究机构之一;该中心在表面活性剂化学的研究方面有长期值得自豪的记录。

　　我很荣幸地接受张高勇院长的邀请,来探索中国日用化学研究院、武汉大学和Bristol大学之间合作的可能性。我们认为开辟对话的一个好办法是我用英语先作场讲座。在动身之前我准备了一个小册子,其中包含帮助理解的教材,也谈到很多较专业化的研究领域。

　　我的讲座计划用计算机多媒体技术进行,为了保证不出技术上的故障,演示内容和讲稿的CD被提前寄出。CD仅在我到太原之前的几天到达,我却惊讶地发现张教授的博士生在如此短的时间内把所有的英语译成了中文!有力地展示了中国学生的无限精力和热情。

　　讲座由老朋友杜志平(中国日化院、Leeds大学)现场翻译。

　　本书采取中英文对照格式:即中文页直接对应于后面的英文页。目的是拓宽胶体与表面科学在中国的感染力,加强科学理解和交流技巧,希望这种并行的英－中格式能帮助吸引更多的学生到该领域。

　　谨对以下参与本书翻译的学生表示谢意:洪昕林、朱银燕、张剑、侯潇云、张越、张杰、薛长晖、陈志萍、姚建国、敬登伟、马利静;最后由周晓海和董金凤统稿。没有你们的努力就没有本书,因此可以说它是你们的作品——祝贺你们!

　　感谢武汉大学研究生院和化学学院对该书出版的大力支持和鼓励。

　　　　　　　　　　　　　Dr. Julian EASTOE　　　　Bristol UK 12 May 2003

序

　　表面活性剂是一类易吸附于界面,对界面性质产生影响的两亲分子化合物。由于这些化合物在溶液中的分子聚集体的粒径为 1~1 000nm,呈胶体性质,因此表面活性剂物理化学成为胶体与界面科学研究领域的主题点和中心内容。胶体与界面科学是一门密切联系生产实际的科学,其研究不仅已成为许多重要材料和工程的理论基础,而且其实际应用已渗透到工农业生产,乃至国民经济的各个领域。自从 20 世纪 80 年代两亲分子有序组合体(organization of amphiphilic molecules)概念的提出,1992 年 Mobil 公司首次用表面活性剂自组装体系作为模板合成了介孔材料以来,随着计算机和现代实验技术的应用,对表面活性剂在界面上的吸附和溶液中的自组装行为从分子水平上的解析,使人们对表面活性剂这类两亲分子化合物有了更深的理解和认识,并将其广泛用于材料、医药、生物等高新技术领域的科学研究中。

　　Bristol 大学是英国最有威望的大学之一,其化学学院在英国高等教育和高校科研评比中一直名列前茅。Bristol 大学对胶体与界面科学的研究有 100 年值得骄傲和丰富的历史,6 位 Leverhulme 教授都已跻身于胶体和界面科学的国际著名人士中。自 1998 年任 Bristol 大学化学学院 Reader 以来,Eastoe 博士一直活跃在胶体与界面科学的研究领域中,在行业知名刊物上发表研究论文 100 多篇;2000 年以来应邀在美国、澳大利亚、日本及欧洲讲学 36 次;自 1995 年以来一直当选为行业导向杂志 Langmuir 的编委;2002 年被聘为中国日化院名誉教授;2003 年获 Kyoto 大学日本促进科学会访问研究员职位。Eastoe 博士在教学上更有其独到之处,受到学生的广泛好评,曾获 Bristol 大学 1999 年优秀教师奖。

　　《表面活性剂化学》的出版,将推动胶体与界面科学与化学各领域及物理学、生命科学和环境科学多学科的交叉。全书分两册,一册主要介绍表面活性剂的基本原理和物化性能,另一册将着重介绍研究表面活性剂的最新方法、研究进展和在交叉学科中的应用。本书是在 Eastoe 博士 2002 年 10 月在中国日化院的讲座稿的基础上修改而成的,由于时间有限,对表面活性剂物理化学不可能进行全面和深入的介绍,但我相信读者会从中获益。另外,它的独特的中英文对照格式,使读者能在汲取科学知识的同时,欣赏到地道的英语表达方式。本书还可作为高等学校中双语教学的参考教材,相信对读者英语交流和化学专业水平的提高都会有所帮助。

2003 年 5 月

编 者 的 话

胶体与界面化学是物理化学的重要分支之一,内容涉及信息、生物、环境、能源等前沿学科,特别是纳米材料的制备、溶液中有序分子组合体、生物膜模拟等课题,使从事胶体与界面化学的研究者应接不暇。

我国多数高等学校里,化学基础课程通常对胶体与界面化学的内容涉及不多,大学化学专业的学生对胶体与界面化学基本原理了解得不够。因此,有必要开设胶体与界面化学课程,让学生了解其基本原理以及现代尖端材料与胶体界面化学的关系,为相关学科的发展服务以及提高学生科学研究能力,这就是我们整理编译此书的目的。

本书是基于英国 Bristol 大学 Julian Eastoe 博士在中国日用化学工业研究院所作讲座的讲稿翻译编撰而成的。在张高勇院士的具体指导下,由武汉大学胶体与界面科学实验室的博士生洪昕林、朱银燕以及中国日用化学工业科学研究院的张剑、侯潇云、张越、张杰、薛长晖、陈志萍、姚建国、敬登伟和马利静等协力完成了该书的翻译,最后由胶体与界面科学实验室的周晓海和董金凤校正和统稿。在编译过程中,得到了武汉大学研究生院和化学学院的大力支持和鼓励,对此,谨致以最诚挚的感谢。

表面活性剂化学以中英文对照的形式出版,以适应双语教学的需要。全书共分上、下两册,上册为胶体与界面化学的基本原理,特别是表面活性剂的溶液特性;下册为表面活性剂化学的最新发展,涉及许多尖端材料的胶体界面化学。由于原文内容丰富,涉及的知识面较宽,受编译人员专业的局限和对原文的理解存在差异,本书难免有不妥与错误之处,敬请指正。

武汉大学胶体与界面化学实验室
2003 年 5 月于武汉

目　录

第一章　表面活性剂化学及其一般相行为 ················· 1
第二章　界面的聚集与吸附 ····························· 8
第三章　微乳液 ······································· 33
第四章　散射技术 ····································· 54
1. Surfactant chemistry and general phase behaviour ······ 76
2. Aggregation and adsorption at interfaces ············· 86
3. Microemulsions ································· 121
4. Scattering techniques ···························· 153

第一章 表面活性剂化学及其一般相行为

1.1 胶体体系中的表面活性剂

胶体(希腊语 glue)这个术语是在 1861 年由 Thomas Graham 首先提出来的，用来描述 Francesco Selmi 在 19 世纪中期制得的 AgCl、硫和普鲁士蓝等"准溶液体系"[1]。这些体系的典型特征是粒子不因重力而沉降且在体系中扩散速率低。Graham 由此推断出胶体粒子的粒径范围大约在 $1\mu m \sim 1nm$ ($10^{-6} \sim 10^{-9}$ m)。该术语仍适用于现代胶体体系，胶体一般是指一种物质以细微状态均匀分散在另一种物质中的体系，分别被称为分散相和分散介质(即连续相)，它们可以是固体、液体或气体。与胶体粒子的特征尺度相关的巨大比表面积将导致产生许多不同的体系以及大量实际应用的发展，并涉及重要的界面现象。

在这些体系中，最常见、最古老的可能是憎液胶体，它由不能溶解或不能混合的成分组成。这可以追溯到 19 世纪 50 年代，Michal Faraday 制得的胶态金溶胶(溶胶涉及固体颗粒在水中的分散)[2]。较常见的憎液胶体有牛奶(脂肪小液滴分散在水相中)、烟(固体颗粒分散在空气中)、雾(小液滴分散在空气中)、涂料(固体小颗粒分散在液体中)、果冻(大蛋白质分子分散在水中)、骨骼(磷酸钙分散在胶原质的固体基质中)。第二种胶体体系则是亲液胶体，这种胶体是自发形成、热力学稳定的溶液体系。这些体系可由聚合物(也就是比溶剂分子大许多倍的溶质分子)的溶液组成，成为一个重要而独特的研究领域(聚合物科学)。

胶体体系的另一种主要类型是所谓的"聚集胶体"，它也归属为亲液性的胶体。即两亲(亲油性和亲水性)分子的聚集体，两亲分子在动力学和热动力学驱动下聚集，这种体系既是分子溶液体系，也是真正的胶体体系。构成此类胶体的分子通常称为"表面活性剂"，即表面活性物质的缩写。第二章将详细介绍表面活性剂这类多功能的重要化合物。由于表面活性剂分子的双亲性，许多重要的界面现象与之相关，例如润湿，同样它们也存在于很多工业产品和工艺过程中。

1.2 表面活性剂的特性

表面活性物质是有机化合物,当在溶剂中的浓度较低时,它们易吸附于界面,从而显著地改变界面的物理性质("界面"一般是指液/液、固/液、气/液界面,下文中也使用了"表面"一词)。这种吸附行为决定于溶剂的性质和表面活性剂的化学结构,表面活性剂分子在其分子结构中同时含有极性和非极性基团(即两亲性质)。基于此,当两亲分子位于界面处时,憎液部分向外伸出溶剂表面,而亲液部分仍保留在溶液中。水是最常用的溶剂,也是工业和科研领域中最常见的液体,因而,表面活性剂两亲部分可以看做是亲水部分和憎水部分,常称为亲水头基和憎水尾基。

吸附与能量变化有关,因为吸附在界面上的表面活性剂分子比体相中的表面活性剂分子自由能要低。因而表面活性剂在界面(液/液或气/液)的聚集是一个自发的过程,并且会导致界面(表面)张力的下降。然而,许多物质都有这样的性质:中长链的醇有表面活性(例如:正己醇和十二醇),但是不能称之为表面活性剂。真正的表面活性剂不仅能够在界面(这里指空气/水或者油/水界面)定向地形成单分子层,更重要的是能够在溶液中形成自组装结构(胶束、囊泡等)。同时它们具有乳化、扩散、润湿、发泡或者去污等性质因而不同于普通的表面活性物质。

表面活性剂的吸附和聚集现象均是基于憎水效应[3],即表面活性剂亲油基有自发逃离水相的倾向。这主要是因为水-水分子间的相互作用要强于水-油间的相互作用。表面活性剂的另一个特点是当表面活性剂的水溶液浓度超过40%时,就会出现由表面活性剂分子组成的大的有序聚集结构,亦即液晶相。

表面活性剂体系相行为及其结构的多样性被广泛应用在许多工业领域,尤其是涉及高比表面的粒子、界面活性的改进以及胶体体系的稳定性时。表面活性剂种类繁多,且不同表面活性剂混合使用时能产生独特的协同作用[4],因此表面活性剂的基础和实际应用始终是研究热点。在此我们不一一赘述表面活性剂的物理性质和有关应用,仅在下文中举若干例子来加以说明。

1.3 表面活性剂的分类及其应用

1.3.1 表面活性剂的分类

表面活性剂的亲水基和亲油基均有许多类型。亲水基可以带电荷,也可以是中性;可以是小基团,也可以是聚合链。亲油基可以是单链的,也可以是双链的、直链的或者是支链的烷基烃类,也可以是氟碳化合物、硅氧烷或芳香烃类。表面活性剂中常见的亲水和亲油基团分别列于表1.1和表1.2中。

因为亲水部分通常通过离子间的相互作用或氢键作用而溶于水,所以最简单的分类一般基于表面活性剂亲水基的种类,另外,根据憎水部分的不同有更细的分类。基本分类如下:

- 阴离子和阳离子表面活性剂:它们溶解在水中成为两种带相反电荷的离子(表面活性剂离子和它的反离子)。
- 非离子表面活性剂:含有不带电极性部分,例如聚氧乙烯基团(—OCH_2CH_2O—)、多羟基基团等。
- 两性离子表面活性剂:同时含有阳离子基团和阴离子基团。

在不断改善表面活性剂性能的过程中,发现了一系列新型结构的表面活性剂,它们具有优异的协同作用以及更好的界面和聚集性质。这些新型表面活性剂在过去的二十年中备受关注,阴阳离子混合型表面活性剂、BOLA型、双子型、聚合物型以及可聚合型的表面活性剂都属于此类[5,6],表1.3中列出了一些典型的例子和它们相应的特性。另外,表面活性剂的生物降解性亦越来越受到重视,特别是个人护理用品和家用洗涤用品中[7],不仅要求有高的生物降解性,而且要求配方中的每种成分都无毒副作用。

表1.1　　市场上常见的表面活性剂亲水性基团

种 类	一般结构
磺酸盐	R—SO_3^- M^+
硫酸盐	R—OSO_3^- M^+
羧酸盐	R—COO^- M^+
磷酸盐	R—OPO_3^- M^+
铵	$R_xH_yN^+X^-$ ($x=1\sim3, y=4-x$)
季铵盐	$R_4N^+X^-$
甜菜碱	$RN^+(CH_3)_2CH_2COO^-$
磺化甜菜碱	$RN^+(CH_3)_2CH_2CH_2SO_3^-$
聚氧乙烯(POE)	R—$OCH_2CH_2(OCH_2CH_2)_n$OH
多羟基化合物	蔗糖、山梨聚糖、甘油、乙烯醇等
多肽	R—NH—CHR—CO—NH—CHR′—CO……—CO_2H
聚缩水甘油	R—$(OCH_2CH[CH_2OH]CH_2)_n$—$OCH_2CH(CH_2OH)CH_2OH$

表 1.2　　　　　市场上常见的表面活性剂憎水性基团

基团	一般结构	
天然脂肪酸	$CH_3(CH_2)_nCH_3$	$n = 12 \sim 18$
石油石蜡	$CH_3(CH_2)_nCH_3$	$n = 8 \sim 20$
石蜡	$CH_3(CH_2)_nCH=CH_2$	$n = 7 \sim 17$
烷基苯	$CH_3(CH_2)_nCH_2-\phi$ 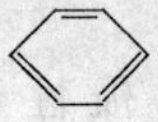	$n = 6 \sim 10$，直链或支链
烷基芳香化合物	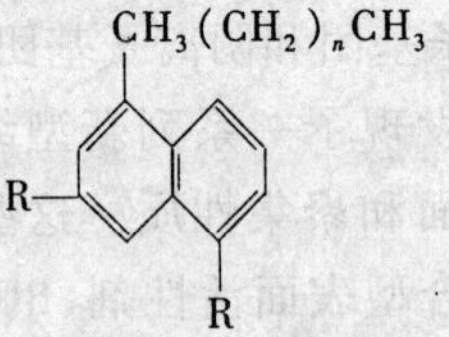 萘环结构，带 $CH_3(CH_2)_nCH_3$ 和 R 取代基	$n = 1 \sim 2$ 为水溶性；$n = 8$ 或 9 为油溶性
烷基苯酚	$CH_3(CH_2)_nCH_2-\phi-OH$ 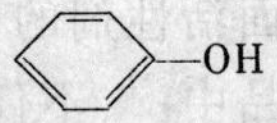	$n = 6 \sim 10$，直链或支链
聚氧丙烯	$CH_3CHCH_2O(CHCH_2)_n$，X 和 CH_3 取代	n 为聚合度；X 为聚合引发剂
碳氟化合物	$CF_3(CF_2)_nCOOH$	$n = 4 \sim 8$，直链或支链，或者终端为氢
硅树脂	$CH_3O(SiO)_nCH_3$，两个 CH_3 取代	

表 1.3　　　　　新型表面活性剂的种类和结构特征

种类	结构特征	示例
阴阳离子混合表面活性剂	阴阳离子表面活性剂体积克分子浓度相等的混合物（无无机反离子）	正十二烷基三甲基铵的正十二烷基硫酸盐（DTADS） $C_{12}H_{25}(CH_3)_3N^+\ ^-O_4SC_{12}H_{25}$

续表

种类	结构特征	示例
BOLA 型表面活性剂	一个聚亚甲基直链上连有两个带正电荷的头基	十六烷-1,16-二(三甲基溴化铵) $Br^-(CH_3)_3N^+—(CH_2)_{16}—N^+(CH_3)_3Br^-$
双子表面活性剂	两个完全相同的表面活性剂连在同一个基团上或者是它们连在同一个头基上	丙烷-1,3-二(十二烷基二甲基溴化铵) C_3H_6-1,3-bis[$(CH_3)_2N^+C_{12}H_{25}Br^-$]
聚合物型表面活性剂	具有表面活性的聚合物	异丁烯和琥珀酸酐的共聚物 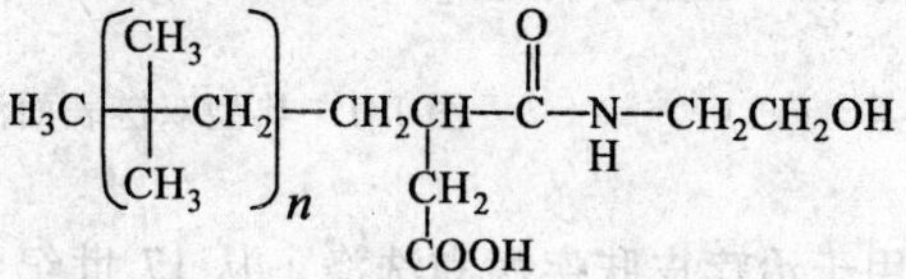
可聚合型表面活性剂	表面活性剂可发生均聚或者和体系中的其他成分发生共聚	11-(丙烯酰)十一烷基三甲基溴化铵 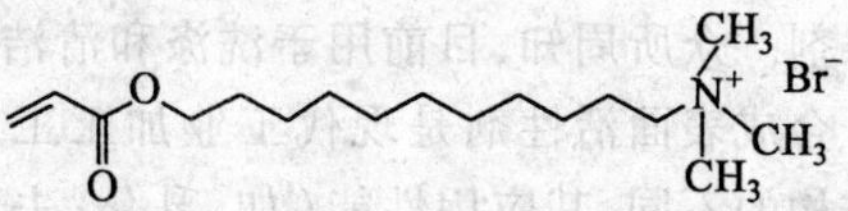

二(2-乙基己基)琥珀酸磺酸钠是典型的双链表面活性剂,商业名称为气溶胶-OT 或 AOT。图 1.1 列出了四个双链表面活性剂的典型实例的化学结构,其中就有 AOT。

1.3.2 表面活性剂的应用与发展

表面活性剂既有天然的,也有人工合成的。天然表面活性剂包括天然生成的双亲分子,如脂质体,它们是以甘油酯为骨架的表面活性剂,是细胞膜的重要成分。常见的肥皂即属于此类[8]。最先认识它们的表面活性可追溯到古埃及时代,当时通过把动物油和植物油同碱性的盐混合制得类似肥皂的物质,这类物

5

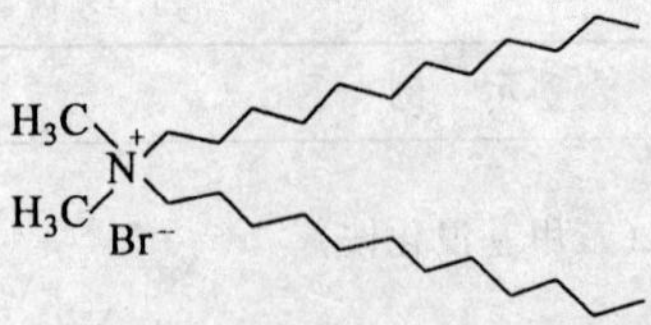

阳离子型：正双十二烷基二甲基溴化胺
(DDAB)

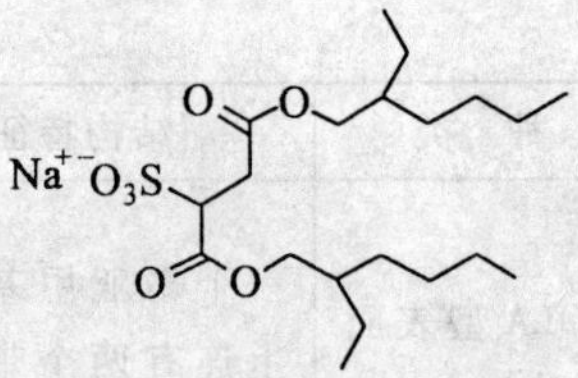

阴离子型：二(2-乙基己基)磺化琥珀酸钠(气溶胶-OT 或 AOT)

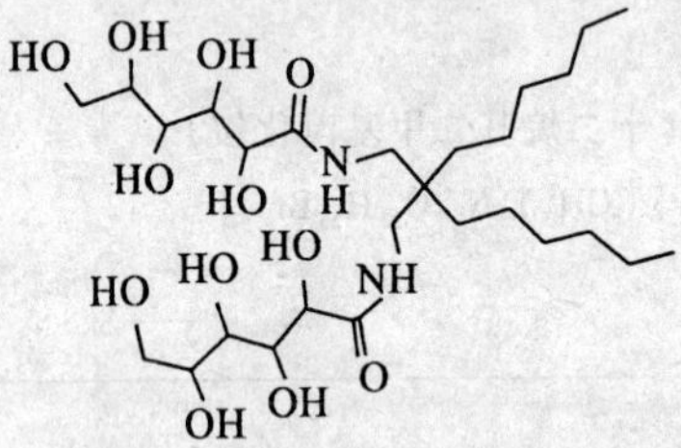

非离子型：二己基葡萄糖胺 [二-(C6-Glu)]

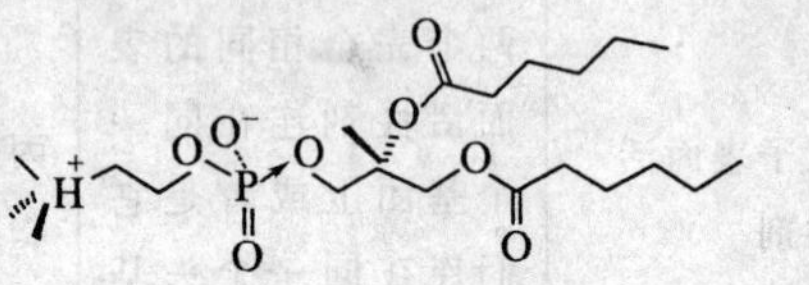

两性离子型：二-己基卵磷脂 [(二 C6)PC]

图 1.1 典型双链表面活性剂的结构

质被用于治疗皮肤疾病和洗涤。从 17 世纪到 20 世纪初期，肥皂仍然是惟一的天然洗涤剂，后来逐渐出现了须用、发用以及沐浴和洗涤用的各类产品。由于第一次世界大战使制皂的原料油脂短缺，德国在 1916 年首先成功制成了人工合成洗涤剂。众所周知，目前用于洗涤和清洁的合成洗涤剂可由许多原材料得到。

合成表面活性剂是现代工业加工工艺和配方中的主要成分[9-11]。由于化学结构的不同，其应用性能(如：乳化、去污和发泡)也是不同的，憎水链的长度和排列以及亲水基团的性质和所处位置决定了表面活性剂分子的性能。通常认为：当链长为 C12 到 C20 时，具有最佳的去污能力；当链长稍短时，则有较好的润湿和发泡作用。结构和性能的关系以及化学相容性是表面活性剂应用中的关键因素，因此该领域一直是研究工作的重点。

在各种表面活性剂中，阴离子表面活性剂的制备工艺简单，且成本低廉，因此用量较大。阴离子表面活性剂含有带负电荷的亲水基，例如肥皂中的羧酸盐($-CO_2^-$)，还有硫酸盐($-OSO_3^-$)和磺酸盐($-SO_3^-$)。它们主要应用于洗涤剂、个人护理用品、乳化剂和肥皂中。

阳离子表面活性剂含有带正电荷的亲水基，例如三甲基季铵离子($-N(CH_3)_3^+$)，它们主要用于带负电荷(如：金属、塑料、矿物、纤维、头发和细胞膜等)吸附质的表面吸附。这些物质也由于阳离子表面活性剂的吸附而改变性能。因此阳离子表面活性剂通常用于抗腐蚀、抗静电、浮选剂、织物柔软剂、发胶和杀菌剂。

非离子表面活性剂的强亲水性来自氢键偶极之间的相互作用,例如乙氧基化物$\{-(OCH_2CH_2)_mOH\}$。非离子表面活性剂较之离子型表面活性剂的优点是,可以同时改变亲水基和憎水基的长度来获得最佳的使用效果。它们可以被用于低温洗涤剂和乳化剂中。

两性离子表面活性剂生产成本高,所以它们是用量最少的品种。它们的特点是有非常好的护肤作用和皮肤协调性。由于对眼睛和皮肤刺激性小,因此常用于香波和化妆品中。

参考文献(略)

第二章 界面的聚集与吸附

由字义来讲,表面活性剂具有表面活性,这是一个包括任何液/液、液/气或液/固体系的广泛领域,本章的主要论述聚焦于表面活性剂在水溶液中的吸附与聚集现象。读者可以阅读一些相关的书籍及专论,以便更好地理解表面活性剂这一概念。

2.1 表面活性剂在界面的吸附

2.1.1 表面张力与表面活性

位于界面的表面活性剂分子与位于体相中的分子所处的环境不同,这可以从表面自由能来理解。例如空气-水界面,由于水分子受到的短程吸引力不平衡,会受到指向体相的净拉力作用。因此,达到与气相最小限度的接触是一自发的过程,这可以用来解释液滴与气泡为什么总是趋于球形这一现象。单位面积的表面自由能,定义为表面张力(γ_0)在每增加单位面积(ΔA)时所需的最小功($W_{\min}$),表示为:$W_{\min} = \gamma_0 \times \Delta A$。还可用另一种不太精确的定义,即拉伸单位长度气-液界面上的液膜所需的力来描述表面张力。

因此,从能量的角度讲,表面活性剂是在低浓度下可在界面上吸附并改变界面能的一类物质。正如第一章所介绍的,表面活性剂的两亲化学性质使其具有明显的降低表面张力的能力。以空气-水界面为例,产生吸附的力来源于体相内的疏水作用。由于水分子之间存在范德华力及氢键,两亲分子通过其疏水基团影响到水所形成的结构,使体系的自由能升高,这就是疏水效应[1]。表面活性剂分子到达界面所需的功要比水分子所需的功小得多,因此表面活性剂向表面聚集是一个自发和优先的过程。其结果是在气-液界面产生新的表面和定向的表面活性剂单分子层。在该表面单分子层中,两亲分子的疏水基团向外,而头基指向内部,即水相。为抵抗在正常表面张力作用下表面收缩的趋势,两亲分子使表面(或膨胀)压力 π 增加,并因此降低溶液表面张力 γ。表面压力定义为 $\pi = \gamma_0 - \gamma$,式中 γ_0 为无两亲分子时空气-水界面的表面张力。

尽管不同结构的表面活性剂分子有不同的吸附浓度范围和速率,但在临界胶束浓度(CMC)以上均会在体相发生聚集。在CMC处,两亲分子在界面达到(接近)最大吸附,同时进一步使表面自由能最小化,分子开始在体相聚集。在CMC之上,体系由吸附在界面的表面活性剂单分子层、游离表面活性剂分子和体相中的表面活性剂胶束组成,且此三相达到平衡。胶束的结构和形成将在2.3节中做详细描述。在CMC以下,表面活性剂分子以相同的速率到达和离开表面,吸附为动态平衡过程。表面活性剂表面浓度的时间平均值可通过热力学方程(见2.1.2)直接或间接地定义及计算。

动态表面张力——这是与平衡表面张力相对应的表面活性剂体系的一个重要性质,在许多重要工业和生物应用中起决定作用[2~5],例如在印刷和衣料印染过程中,由于新界面不断地产生,致使表面张力无法达到平衡,任何表面活性剂溶液中的平衡都不会立刻达到,而是经由表面活性剂分子首先由体相扩散到表面,然后在界面上吸附和定向排列的过程。因此,表面活性剂溶液中新鲜界面的表面张力与溶剂的表面张力非常接近,随后此动态表面张力会在一定时间内达到一平衡值。基于表面活性剂的类型和浓度不同,达到这一平衡的过程从几毫秒到几天。为了能对表面活性剂的这一动态性质进行控制,充分了解支配表面活性剂分子由体相向界面运动时的主要过程是非常必要的。这一领域的研究已经受到广泛关注,最近的发展动态见参考文献[6~8]。但在目前研究工作中一般所指的表面张力是平衡表面张力。

2.1.2 表面超量与热力学吸附

定向单分子层的形成引出了一个相关的基本物理量,即表面超量。其定义为表面活性剂分子在表面层中的浓度与体相浓度之差。Gibbs率先给出了由组分导致体系表面张力变化的通用热力学处理方法[9]。

Gibbs吸附方程的一个重要近似是对界面的"确切"定位。在表面活性剂水相(α)与水蒸气相(β)达到相平衡的前提下,界面是一厚度为τ的模糊区域,其性质由α相到β相不断变化。由于真实界面的性质不好确定,为方便起见假设以一个厚度为0的数学平面为界面,在特定值X处的该界面上下两相被认为分别具有与α和β相同的性质。图2.1为这一理想体系的示意图。

在Gibbs划分面定义中,Gibbs平面的位置X是以该处溶剂的表面过剩量为0确定的。所以溶液组分i的表面过剩量可表示为

$$\Gamma_i^\sigma = \frac{n_i^\sigma}{A} \qquad (2.1.1)$$

式中,A为界面面积,如果体相中α相和β相延伸至XX'面,则n_i^σ为溶质组分i

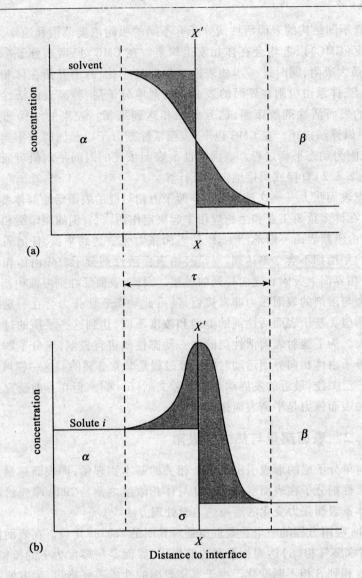

(a) Gibbs 平面为中溶剂过剩浓度为 0 的平面(图中平面两侧的阴影区面积相等);(b) 组分 i 的表面过剩即为该组分在 Gibbs 平面两侧组分浓度(阴影区)的差值

图 2.1 在 Gibbs 划面法中,定义表面过剩浓度为 Γ

在表面相 σ 中相比于等量体相的过剩量。Γ_i^σ 可正可负,其值依赖于 XX' 面的位置。

溶质在包括 α 相和 β 相在内的整个体系的内能为

$$U = U^\alpha + U^\beta + U^\sigma$$

$$U^\alpha = TS^\alpha - PV^\alpha + \sum_i \mu_i n_i^\alpha \tag{2.1.2}$$

$$U^\beta = TS^\beta - PV^\beta + \sum_i \mu_i n_i^\beta$$

界面相 σ 对应的热力学内能表达为

$$U^\sigma = TS^\sigma + \gamma A + \sum_i \mu_i n_i^\sigma \tag{2.1.3}$$

当 T,S,A,μ,n 变化为无穷小时，(2.1.3)式变化为

$$dU^\sigma = TdS^\sigma + S^\sigma dT + \gamma dA + Ad\gamma + \sum_i \mu_i dn_i^\sigma + \sum_i n_i^\sigma d\mu_i \tag{2.1.4}$$

在小的可逆变化过程中，体相内任一组分的内能微分形式为

$$dU = TdS - PdV + \sum_i \mu_i dn_i \tag{2.1.5}$$

类似地，界面区的内能微分形式为

$$dU^\sigma = TdS^\sigma + \gamma dA + \sum_i \mu_i dn_i^\sigma \tag{2.1.6}$$

(2.1.4)式减去(2.1.6)式，得

$$S^\sigma dT + Ad\gamma + \sum_i n_i^\sigma d\mu_i = 0 \tag{2.1.7}$$

在恒定温度下，组分 i 的表面超量如(2.1.9)式所示为 Γ_i^σ，则 Gibbs 方程的通用形式为

$$d\gamma = -\sum_i \Gamma_i^\sigma d\mu_i \tag{2.1.8}$$

对于由溶剂和溶质组成的简单体系，分别以下标 1 和 2 代表溶质和溶剂，则(2.1.8)式表示为

$$d\gamma = -\Gamma_1^\sigma d\mu_1 - \Gamma_2^\sigma d\mu_2 \tag{2.1.9}$$

由于在选择 Gibbs 划分面时，取 $\Gamma_1^\sigma = 0$，则(2.1.9)式简化为

$$d\gamma = -\Gamma_2^\sigma d\mu_2 \tag{2.1.10}$$

式中，Γ_2^σ 为溶质的表面过剩浓度。

其中化学势为

$$\mu_i = \mu_i^o + RT\ln a_i \quad \therefore \quad d\mu_i = \text{cste} + RTd\ln a_i \tag{2.1.11}$$

式中，μ_i^o 代表组分 i 在 1atm 和 298K 时的标准化学势。

对于非解离性物质(如非离子表面活性剂)应用(2.1.10)式得出 Gibbs 方程的通用形式为

$$d\gamma = -\Gamma_2^\sigma RTd\ln a_2 \tag{2.1.12}$$

或
$$\Gamma_2^\sigma = -\frac{1}{RT}\frac{d\gamma}{d\ln a_2} \tag{2.1.13}$$

对解离性溶质（如 R^-M^+ 型的离子型表面活性剂），假设当其低于 CMC 时为理想状态，(2.1.12)式变为

$$d\gamma = -\Gamma_R^\sigma d\mu_R - \Gamma_M^\sigma d\mu_M \tag{2.1.14}$$

在不加电解质时，界面的电中性要求 $\Gamma_R^\sigma = \Gamma_M^\sigma$，引入平均离子活度 $a_2 = (a_R a_M)^{1/2}$，代入(2.1.14)式，即得 1:1 型解离溶质的 Gibbs 方程：

$$\Gamma_2^\sigma = -\frac{1}{2RT}\frac{d\gamma}{d\ln a_2} \tag{2.1.15}$$

在加入过量电解质（即盐量足以削弱静电效应）且电解质的反离子 M^+ 与表面活性剂的反离子相同的条件下，则 M^+ 的活度为常数，得出 Gibbs 方程中的前因子为 1，此时得到的方程与(2.1.13)式相同，所以(2.1.13)式在此条件下仍然适用。

对于表面活性剂等在界面有强烈吸附的物质，微小的体相浓度变化即可引起界面张力（表面张力）的较大变化。通过测量界面张力和溶质浓度的关系可以得到溶质在界面的相对吸附量及表面活性等性质。值得注意的是：对于稀表面活性剂体系，一般情况下在使用(2.1.13)式和(2.1.15)式时可用活度代替浓度，对精度影响不大。

图 2.2 给出了随表面活性剂浓度的增加，水的表面张力典型降低的过程，同时给出了 Gibbs 方程(2.1.13)式或(2.1.15)式如何确定表面吸附量。在低浓度区，随着组分 2 表面超量的增加（由 A 区向 B 区），可观察到体系表面张力的降低（25℃时纯水的表面张力为 72.5mN·m^{-1}）。在浓度接近 CMC 时，表面张力趋于一个定值，此时表面张力曲线基本为直线（B 区到 C 区）。

最近对离子表面活性剂的 Gibbs 方程中前因子项的讨论较多（例如参考文献 10~13）。所提出的主要问题是关于离子表面活性剂能否完全解离，如果表面活性剂不能完全解离，而是在其表面附近存在一个过渡层，所得出的前因子值会低于 2。最近通过结合表面张力以及中子反射等能够直接测定表面超量的测量技术，证实了对于离子表面活性剂采用前因子值为 2 是合理的[14]。

尽管在液-液、液-气界面吸附中 Gibbs 等式是最普遍、最常用的数学关系式，但其他情况下一些等温吸附则可以用 Langmuir[15], Szyszkowski[16], Frumkin[17] 方程来表示。Guggenheim 和 Adam 通过不同的分平面法将界面区划分成单独一相（即界面相，且体积有限）[18]，从而简化了 Gibbs 等式。

2.1.3 表面活性剂吸附的效率和效能

表面活性剂降低溶液表面张力的性能可以通过以下两方面来衡量：(1)溶

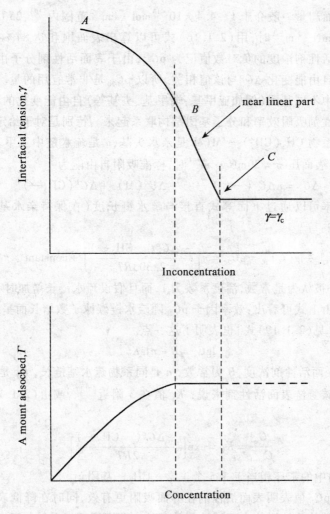

图2.2 由表面张力测量及 Gibbs 吸附方程确定界面吸附等温线

液降低到指定表面张力时所需要的表面活性剂浓度。(2)该表面活性剂所能产生的最低表面张力,即表面活性剂的效率和效能问题。

测定表面活性剂吸附效率的一个很好的方法就是测定表面张力下降 20 mN·m^{-1} 所需要的表面活性剂浓度。此值接近在界面产生最大吸附所需的最小浓度。Frumkin 吸附等式给出了证明[见(2.1.16)式],该等式将表面张力(或表面压 π)的下降和表面超量联系起来。

$$\gamma_0 - \gamma = \pi = -2.303RT\Gamma_m \lg\left(1 - \frac{\Gamma_1}{\Gamma_m}\right) \qquad (2.1.16)$$

最大表面超量一般介于 $1 \sim 4.4 \times 10^{-10} \mathrm{mol \cdot cm^{-2}}$ 范围内[19]:25℃下当表面张力下降 $20\mathrm{mN \cdot m^{-1}}$ 时,用(2.1.16)式可以算出表面饱和达 84% ~ 99.9%。对应的表面活性剂浓度的负对数值记为 $\mathrm{p}C_{20}$,由于表面活性剂分子由体相到界面所产生的自由能变化 $\Delta G°$ 与该值相关,所以 $\mathrm{p}C_{20}$ 是非常有用的量。通过表面活性剂分子中不同基团(例如亚甲基、端甲基、头基等)自由能变化的贡献,就可以将表面活性剂吸附效率和分子基团结构联系起来。特别是对于给定的直链表面活性剂同系物 $\mathrm{CH_3(CH_2)_\mathit{n}}$—M(M 是亲水头基,$n$ 是疏水链中亚甲基数)的水溶液,当体系表面压 $\pi = 20\mathrm{mN \cdot m^{-1}}$ 时,标准吸附自由能为

$$\Delta G° = n\Delta G°(—\mathrm{CH_2}—) + \Delta G°(\mathrm{M}) + \Delta G°(\mathrm{CH_3}—) \quad (2.1.17)$$

吸附效率可以通过下面等式直接与疏水链长度(在保持亲水基不变情况下)关联起来:

$$-\lg(C)_{20} = \mathrm{p}C_{20} = n\left[\frac{-\Delta G°(—\mathrm{CH_2}—)}{2.303RT}\right] + \mathrm{constant} \quad (2.1.18)$$

式中,$G°(\mathrm{M})$ 被认为是常数,活度系数为 1,而且假设疏水链长增加时 $\varGamma_\mathrm{m}$ 不发生明显变化。由上式可看出:效率因子 $\mathrm{p}C_{20}$ 随疏水链碳原子数增长而呈线性增长。Traube 规则[见(2.1.19)式]也表明了这一点。

$$\lg C_s = B - n\lg K_\mathrm{T} \quad (2.1.19)$$

式中,C_s 是表面活性剂浓度,B 为常数,n 是同系物疏水基链长,K_T 是 Traube 常数。对于直碳氢链表面活性剂来说,K_T 值在 3 附近[21],或由(2.1.18)式类推可得

$$\frac{C_n}{C_{n+1}} = K_\mathrm{T} = \exp\left[\frac{-\Delta G°(—\mathrm{CH_2}—)}{2RT}\right] \quad (2.1.20)$$

此外,化合物中的苯环相当于 3.5 个正常—$\mathrm{CH_2}$—基团。

较大的 $\mathrm{p}C_{20}$ 值表明表面活性剂在界面吸附更有效,同时在降低表面张力上效率更高。其他能够增加表面活性剂效率的主要因素可以总结如下:

- 在相同碳原子数下,直链烷基疏水基比支链烷基效率更高。
- 单亲水基位于疏水基团尾部时比一个(或多个)位于疏水基团中部的亲水基效率高。
- 非离子或两性表面活性剂比离子表面活性剂效率高。
- 对离子表面活性剂而言,通过以下两种方法都可以使极性基团的有效电荷减少,提高吸附效率:(a)用束缚能力更大(水化能力差)的反离子;(b)增加水相中的离子强度。

采用表面张力降低 $20\mathrm{mN \cdot m^{-1}}$ 作为衡量表面活性剂吸附效率的标准值虽然方便,但却有一些随意性。此外,当体系内表面活性剂的最大表面超量明显不

同或体系表面压低于 20 mN·m^{-1}时,上述标准值就不适用了。Pitt 等[22]通过定义 $\Delta\gamma$ 为体系在 CMC 时表面压的一半来解决上述问题。

表面活性剂的性能也可以通过吸附效能来讨论。一般是不考虑表面活性剂浓度,而由表面活性剂所能达到的最小表面张力 γ_{min} 或用界面达到饱和时(即达到最大吸附量 Γ_m 时)的表面超量浓度来衡量。Γ_m 和 γ_{min} 主要由临界胶束浓度决定,而对于某些离子表面活性剂的 Γ_m 和 γ_{min},则主要由溶解度上的限制以及 Krafft 点决定,在 2.2.1 部分将进一步对此进行描述。通过 Gibbs 吸附等式,Γ_m 能给出界面堆积情况,所以吸附效能对于决定表面活性剂的发泡性、润湿性、乳化性等特性方面非常重要。

正如 Rosen 所得的大量数据显示的那样,表面活性剂的吸附效率和吸附效能并不一致,表面活性剂常常在低浓度下能显著降低表面张力(即吸附效率高),但 Γ_m 值却很低(即对应吸附效能低)。表面活性剂的分子结构对吸附效率的影响基本只在热力学范畴起作用,而吸附膜中表面活性剂亲水基和疏水基的相对大小却直接与吸附效能有关。每个表面活性剂分子所占据的面积既由疏水链横截面积决定,也由亲水头基达到最紧密堆积所需面积决定,而不考虑哪种效应更大。因此,表面活性剂膜的紧密或松散堆积导致了界面性质的很大差异。例如,直链且具有较大亲水基(相对于疏水链横截面)的表面活性剂倾向于形成紧密、高效的界面堆积。而支链、多链或疏水基较大则阻止分子在界面形成有效紧密堆积。此外,在单直链表面活性剂体系中,从 C_8 到 C_{20} 增加碳氢链长度几乎对吸附效能没有影响[19]。

2.2 表面活性剂的溶解性

在表面活性剂水溶液中,当所有界面均被饱和后,表面活性剂分子通过聚集方式继续降低体系能量。根据体系的组成不同,表面活性剂分子在聚集形式(包括胶束、液晶、双层结构或囊泡等)中所起的作用不同。这种机制的物理表现形式就是表面活性剂可以由溶液中结晶或沉淀,即与体相分离。通常大多数表面活性剂在水中有足够的溶解性,但其溶解性受疏水基长度、亲水基性质、反离子的化合价、溶液环境的影响,尤其是受温度等的影响较显著。

2.2.1 Krafft 温度

大多数溶质在水中的溶解度会随温度的升高而增加。但是,对低温不溶的离子型表面活性剂而言,常常在达到某一温度后,其溶解度会突然增大,这就是所谓的 Krafft 点或 Krafft 温度,即 T_K,其定义为溶解度曲线和 CMC 曲线的交叉点。即在 T_K 时,表面活性剂单体的溶解度与其 CMC 相等,见图 2.3。在 T_K 点以下,表面活性剂单体仅与其水合晶体平衡存在,但在 T_K 之上,由于胶束的形成,大大增加了其溶解性。

离子表面活性剂的 Krafft 点随反离子、烷烃链长度、链结构的不同而不同。对 Krafft 点的认识在许多应用中是至关重要的,这是因为在 Krafft 点以下,表面活性剂不能充分体现其作用,也没有临界表面张力和胶束形成等典型特征。通常可以通过引入长的支链、不饱和键或大的亲水基而得到具有较低 Krafft 点而且能有效降低表面张力的表面活性剂,因为这些基团的引入可以降低分子间的形成结晶的作用力。

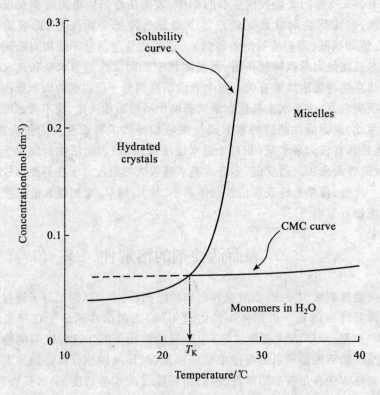

图 2.3 Krafft 温度(T_K)为表面活性剂的溶解度与临界胶束浓度相等的点
(在 T_K 之上,表面活性剂分子形成分散相;而在 T_K 以下,以水合晶体的形式存在)

2.2.2 浊　点

对于非离子表面活性剂,常常发现在某一特定的温度下,透明胶束溶液会明显浑浊,这就是通常所说的浊点。当温度高于浊点时,表面活性剂体系发生相分离。体系由浓度等于CMC的无胶束稀溶液和含大量表面活性剂的胶束相组成。由于聚集数的快速增大和胶束间排斥力的降低[24,25],引起胶束聚集相与无胶束相密度产生较大差异,从而使体系发生相分离。由于形成的大胶束对光有强烈的散射作用,使溶液变得明显浑浊。与Krafft温度相似,浊点也受化学结构的影响。对聚氧乙烯(POE)型非离子表面活性剂而言,当疏水基团不变时,浊点随分子中OE含量的增加而升高。对于OE含量相同的表面活性剂,可以通过减小其疏水基尺寸、加宽POE链长分布以及对疏水基支链改变来降低浊点。

2.3 胶束化

当表面活性剂浓度足够高时,除了能在界面形成定向单分子层外,表面活性剂还能聚集成胶束形态。典型的胶束由50～100个表面活性剂分子组成,胶束的大小和形状由几何因素和能量因素来决定。胶束是在一个非常窄的浓度区域内形成的,这一区域称做临界胶束浓度(CMC)。当浓度大于CMC时,游离的表面活性剂单分子浓度保持不变,多余的表面活性剂形成胶束。因此,表面活性剂溶液的许多平衡或动态性质均在该浓度范围内发生突变(见图2.4)。

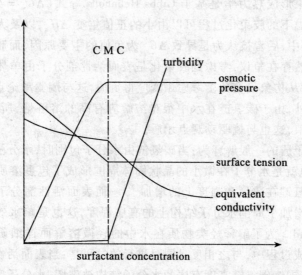

图2.4 可形成胶束表面活性剂溶液的一些物理特性随浓度变化的示意图

2.3.1 胶束化过程的热动力学

胶束是一个动态的概念,也就是说胶束是分子以微秒级时间单位在聚集相和溶液相中不断高速交换的结果。这种不断的形成-解离过程依赖于体系中各种相互作用力的一种微妙平衡。体系中的相互作用主要来自于:(1)碳氢链-水相互作用;(2)碳氢链-碳氢链相互作用;(3)亲水基-亲水基相互作用;(4)亲水基的溶剂化作用。因此,胶束化过程的净自由能变化 ΔG_m 可以用下式表示:

$$\Delta G_m = \Delta G(\text{HC}) + \Delta G(\text{contact}) + \Delta G(\text{packing}) + \Delta G(\text{HG}) \quad (2.3.1)$$

式中:
- $\Delta G(\text{HC})$ 是碳氢链由水相到胶束油性内核的自由能变化。
- $\Delta G(\text{contact})$ 是由溶剂与碳氢链在胶束中接触而引起的表面自由能变化。
- $\Delta G(\text{packing})$ 是由碳氢链被局限于胶束内核而引起的正的自由能变化。
- $\Delta G(\text{HG})$ 是由静电效应以及亲水基的空间构象效应等相互作用所引起的正的自由能变化。

表面活性剂分子聚集的部分原因是由于分子憎水基团通过在溶剂中形成油性微区而趋于尽量减少与水的接触引起的。在溶剂内形成的油性微区中烃链间的相互作用是主要的相互作用,而亲水基团却仍由水包围。

传统的胶束形成热力学是基于 Gibbs-Helmholtz 等式($\Delta G_m = \Delta H_m - T\Delta S_m$)而提出的。室温下的胶束化过程可以由小的正值焓变 ΔH_m,以及大的正值熵变 ΔS_m 来表征,其中,后者被认为是导致 ΔG_m 为负值的主要原因,而胶束化是否是熵驱动过程仍然存在争议,考虑到胶束化是表面活性剂分子由单体到有序聚集体的变化过程,从构象熵变角度考虑熵变应得负值,这与熵驱动论点中大的正熵变相矛盾。此外,由于碳氢链在水中很难溶解应有高的溶解焓,所以 ΔH_m 应该是一个较大的值,这也与熵驱动观点矛盾。

对于这些矛盾的一般解释为:当碳氢链由水分子包围时,水分子形成一种笼状空腔结构(也就是水分子在微小的晶状固体周围形成了包裹溶质的笼子),因此导致了有效氢键在数量和强度上的增加[27]。而表面活性剂分子中碳氢链的主要作用就是增加了周围水分子结构上的有序程度,这也是疏水效应的一个主要表现。Tanford[1]为了解释烃类物质在水中的轻微溶解而详细研究了疏水效应。在胶束形成过程中,与之相反且同时进行的过程是:当表面活性剂憎水部分形成聚集体时,围绕碳氢链高度有序的水分子结构被破坏,水分子又回到其正常状态,所以整个过程是熵增加过程。其他一些研究者也引用了水结构的概念解释上述矛盾[28,29]。

然而,上述论点并不能解释最近在较高温度(最高达166℃)的水体系以及以肼类物质为溶剂的体系中的研究发现[30]。因为在这些体系中,水分子不具备其多数的特定结构性质,因此不可能在疏水物质周围形成高度有序结构的水。

表面活性剂单体分子 S 的胶束化机理可以用下列等式表示:

$$S + S \xrightleftharpoons{K_2} S_2 + S \xrightleftharpoons{K_3} S_3 \cdots \xrightleftharpoons{K_n} S_n + S \rightleftharpoons \cdots \quad (2.3.2)$$

式中,平衡常数 K_n 中 $n = 2 \sim \infty$,聚集过程中不同的热力学参数($\Delta G°$, $\Delta H°$, $\Delta S°$)也可以用平衡常数 K_n 表示。但每个 K_n 不能被单独测量,所以必须通过一系列近似得到这一自组装过程的热力学模型。有两个不精确但常见的模型:质量作用模型和相分离模型。在质量作用模型中,考虑到在 CMC 附近的球状胶束大小非常有限,因此假设 K_n 值中只有一个是主要的,胶束和单体分子有如下化学平衡:

$$nS \rightleftharpoons S_n \quad (2.3.3)$$

式中,n 是形成胶束的表面活性剂分子 S 的个数,即胶束聚集数。在相分离模型中,胶束被认为是体系在浓度大于 CMC 时形成的一个新相。用下式表示:

$$nS \rightleftharpoons mS + S_n \quad (2.3.4)$$

式中,m 是在溶液中游离状态的表面活性剂分子个数,S_n 为新相。在上述两种模型中,单体分子和表面活性剂胶束之间的平衡都由对应的平衡常数 K_m 表示,即

$$K_m = \frac{[\text{micelles}]}{[\text{monomers}]^n} = \frac{[S_n]}{[S]^n} \quad (2.3.5)$$

式中,括号项为相应的摩尔浓度,n 为胶束所含表面活性剂分子单体的个数(聚集数)。尽管胶束化本身是非理想过程[31,32],(2.3.5)式仍假设可用浓度代替活度。

由(2.3.5)式推导,胶束化过程中每摩尔胶束的标准自由能为

$$\Delta G°_m = -RT\ln K_m = -RT\ln[S_n] + nRT\ln[S] \quad (2.3.6)$$

而每摩尔表面活性剂分子的标准自由能变化为

$$\frac{\Delta G°_m}{n} = -\frac{RT}{n}\ln[S_n] + RT\ln[S] \quad (2.3.7)$$

在 CMC 或 CMC 附近,$[S] \approx [S_n]$,所以(2.3.7)式右边第一项可以忽略,每摩尔非离子表面活性剂分子的标准自由能变化可以近似表示为

$$\Delta G°_{M,m} \approx RT\ln(\text{CMC}) \quad (2.3.8)$$

对于离子表面活性剂,由于相应的反离子的存在,必须考虑反离子与表面活性剂单体分子以及胶束的结合度,其质量作用等式为

$$nS^x + (n-p)C' \rightleftharpoons S_n^\alpha \quad (2.3.9)$$

式中，C 为游离状态反离子的浓度，表面活性剂分子在胶束中的离解度 α 由等式 $\alpha = p/n$ 给出。

对于离子表面活性剂，(2.3.5)式可表示为

$$K_m = \frac{[S_n]}{[S^x]^n \times [C^y]^{(n-p)}} \quad (2.3.10)$$

式中，p 为参与胶束形成，但未被胶束束缚的反离子浓度。因此，胶束形成的标准自由能可以用下式表示：

$$\Delta G_m^\circ = -RT\{\ln[S_n] - n\ln[S^x] - (n-p)\ln[C^y]\} \quad (2.3.11)$$

在 CMC 附近，对于完全离子化的表面活性剂有：$[S^{-(+)}] = [C^{+(-)}] = $ CMC，由此近似可得出每摩尔表面活性剂的标准自由能变化，即

$$\Delta G_{M,m}^\circ \approx RT\left(2 - \frac{p}{n}\right)\ln(\text{CMC}) \quad (2.3.12)$$

在高电解质浓度时，(2.3.12)式可以简化为应用于非离子表面活性剂的 (2.3.8)式。

由 Gibbs 函数以及热力学第二定律，非离子表面活性剂的 ΔS° 可以表示如下：

$$\Delta S^\circ = -\frac{d(\Delta G^\circ)}{dT} = -RT\frac{d\ln(\text{CMC})}{dT} - R\ln(\text{CMC}) \quad (2.3.13)$$

由 Gibbs 函数以及(2.3.8)式和(2.3.13)式，非离子表面活性剂形成胶束的焓变 ΔH° 为

$$\Delta H^\circ = \Delta G^\circ + T\Delta S^\circ = -RT^2\frac{d\ln(\text{CMC})}{dT} \quad (2.3.14)$$

同样，对于离子表面活性剂也可表示为

$$\Delta H^\circ = -RT^2\left(2 - \frac{p}{n}\right)\frac{d\ln(\text{CMC})}{dT} \quad (2.3.15)$$

相分离模型和质量作用模型都有不足之处。首先是有关活度系数的假设：相比于在稀溶液体系中表面活性剂单体，不应将较大粒径和较多电荷的胶束体系当成理想体系。但上述模型对目前研究的一般体系仍然适用。另一缺陷就是假设胶束是单分散的，多平衡模型的提出弥补了这一缺陷。多平衡模型是质量作用模型的扩展，该模型考虑到计算胶束聚集数的分布函数。该模型的详细导出可参见文献[33-35]。

2.3.2 影响 CMC 的因素

下面将讨论显著影响 CMC 的因素：首先，最主要的因素是表面活性剂的结构（下文将予以介绍）。其次，反离子的性质、添加剂的存在、温度改变等因素对

CMC 的影响也很重要,在讨论 CMC 时应该考虑到这些因素。

1. 憎水基团

碳氢链长度是决定 CMC 的一个主要因素。对于单碳链表面活性剂同系物而言,CMC 随疏水基团链碳原子数增长而呈对数形式降低。关系式一般符合 Klevens 等式[36]:

$$\lg_0(CMC) = A - Bn_c \tag{2.3.16}$$

式中,A 和 B 在确定的同系物和温度下为常数,n_c 是碳氢链 C_nH_{2n+1} 的碳原子数。常数 A 随憎水基团的个数和性质而改变,而对于单离子亲水基的所有链烷烃盐类表面活性剂来说,B 是近似等于 $\lg 2$ ($B \approx 0.29 \sim 0.30$) 的常数(例如,每增加一个—CH_2—,CMC 约降低一半)。

有趣的是,(2.3.16)式仍然适用于直链双烷基磺基琥珀酸酯[37],而且 $B \approx 0.62$,近似等于单链化合物 B 值的两倍。在憎水基团中烷烃链的支链以及双键、芳香基团以及其他极性基团都对 CMC 有显著的影响。在碳氢类表面活性剂中,憎水基团的支链会导致 CMC 升高[19],而链上接入一个苯环相当于增加 3.5 个碳原子。

2. 亲水基团

对于含有同样碳氢链的表面活性剂,亲水性质的不同(如离子表面活性剂与非离子表面活性剂)对 CMC 值也有大的影响。例如,C_{12} 烃链的离子表面活性剂的 CMC 在 1×10^{-3} mol·dm^{-3} 范围内,而相同憎水烃链的非离子表面活性剂的 CMC 在 1×10^{-4} mol·dm^{-3} 范围内。由于胶束形成的主要驱动力是上面讨论到的熵驱动,所以离子基团的性质对 CMC 没有显著影响。

3. 反离子影响

离子表面活性剂胶束的形成与溶剂和离子极性头基间的相互作用密切相关。对于完全离子化的离子表面活性剂极性头基而言,亲水头基之间的静电斥力很大,所以增大离子束缚程度会导致 CMC 降低。对于一个给定憎水链和阴离子亲水头基的表面活性剂,CMC 以 $Li^+ > Na^+ > K^+ > Cs^+ > N(CH_3)_4^+ > N(CH_2CH_3)_4^+ > Ca^{2+} \approx Mg^{2+}$ 顺序降低。对于如十二烷基三甲基卤化铵等阳离子表面活性剂而言,CMC 随 $F^- > Cl^- > Br^- > I^-$ 的顺序降低。此外,反离子电荷的不同对 CMC 也有显著影响。将反离子电荷由一价变为二价、三价会使 CMC 发生突降。

4. 加盐影响

加入惰性电解质会使大多数表面活性剂 CMC 降低,而盐对离子表面活性剂的影响最大。加盐的主要影响是部分屏蔽了极性头基之间的静电斥力,从而降低 CMC。对于离子表面活性剂来说,电解质的影响可由以下经验式定量表示:

$$\lg(\text{CMC}) = -a\lg C_i + b \qquad (2.3.17)$$

加盐对非离子表面活性剂和两性表面活性剂也有类似影响,但(2.3.17)式却不适用。

5. 温度影响

温度对胶束形成过程的影响一般较弱,反映在结合力、热容、体积等的微小变化上。但温度对 CMC 的影响却十分复杂。例如,在温度由 0℃ 升高到 70℃ 时,大多数离子表面活性剂的 CMC 都会出现一个最小值[38]。正如在 2.2 部分所讨论过的,温度的主要影响表现为 Krafft 点和浊点。

2.3.3 胶束结构和分子堆积

早期研究[39,40]表明:单烷基憎水链离子表面活性剂形成球状胶束。最具代表性的是 Hartley[41] 在 1936 年对球状胶束的描述。他认为离子表面活性剂烷基链形成类似液态烃的球状胶束的内核,而极性基团形成了带电的表面。后来,由于两性、非离子表面活性剂的出现,发现胶束可以有不同的形状。胶束在几何形状上的不同被认为主要由表面活性剂的结构以及环境因素(例如浓度、温度、pH 值、电解质含量等)决定。

在胶束形成过程中,分子的几何形状不但起着重要的作用,而且能够帮助人们理解表面活性剂的堆积方式。表面活性剂在溶液中主要的聚集方式有:球状胶束、囊泡、双层结构和反相胶束等。正如前面讨论的那样,表面活性剂自组合过程是由以下两种相反作用所控制:碳氢链与水之间的相互作用和亲水基之间的相互作用。前者迫使表面活性剂分子脱离水环境,促使胶束形成,而后者却起相反作用。这两种作用可以分别被认为是碳氢链引起的界面张力中的吸引作用和由亲水基团引起的界面张力中的斥力作用。最近,Mitchell 和 Ninham[42] 以及 Israelachvili[43] 重新阐述了这个基本观点并将之定量化,最终的结论是:表面活性剂的聚集是由平衡的分子几何形状控制。简而言之,总体自由能与分子几何形状有密切联系(见图 2.5),主要考虑三个关键几何特征:

- 亲水基所占据的最小界面面积,a_0;
- 憎水基(s)的体积,V;
- 胶束内核憎水链最大伸展长度,l_c。

形成球状胶束要求 l_c 等于或小于胶束内核半径 R_{mic}。因此,对于球状胶束的聚集数 N 就可以通过胶束内核 V_{mic} 和表面活性剂憎水链体积 V 以下式表示:

$$N = V_{mic}/v = [(4/3)\pi R_{mic}^3]/V \qquad (2.3.18)$$

或者以胶束面积 A_{mic} 和表面活性剂分子横截面积 a_0 以下式表示:

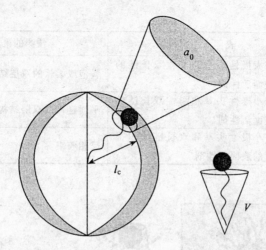

图 2.5 临界堆积因子 P_c（或表面活性剂数）把表面活性剂憎水部分的伸展长度、体积以及亲水基面积关联成无因次量 $P_c = V/a_o l_c$

$$N = A_{mic}/a_o = [4\pi R_{mic}^2]/a_o \qquad (2.3.19)$$

将(2.3.18)式和(2.3.19)式联立，则推出

$$V/(a_o R_{mic}) = 1/3 \qquad (2.3.20)$$

对于球形胶束，l_c 不可能大于 R_{mic}，所以有

$$V/(a_o l_c) \leqslant 1/3 \qquad (2.3.21)$$

一般来说，可以将临界堆积因子 P_c 定义为体积与表面比：

$$P_c = V/(a_o l_c) \qquad (2.3.22)$$

式中，参数 v 随憎水基团的碳链长度、链不饱和度、支链情况以及其他相容性憎水基团的作用变化而变化，而 a_o 主要是由静电作用和头基的水化作用决定。由于通过 P_c 能预测聚集体的形状和尺寸，所以 P_c 是一个非常有用的性质。在整个几何可能范围内的表面活性剂可预测的聚集体特征见表 2.1 和图 2.6

表 2.1　与表面活性剂临界堆积因子, $P_c = V/a_o l_c$ 相关的聚集体特征

P_c	表面活性剂特征	预测的聚集结构
<0.33	单碳链大极性头基的表面活性剂	球形或椭球形
0.33~0.5	单碳链小极性头基的表面活性剂，或存在大量电解质下的离子表面活性剂	大的圆柱状或棒状胶束

续表

P_c	表面活性剂特征	预测的聚集结构
0.5~1.0	有大极性头基和柔软双长链的表面活性剂	囊泡或柔软的双层结构
1.0	有小极性头基和刚性双长链的表面活性剂	平面延伸的双层结构
>1.0	有小极性头基和双长链体积很大的表面活性剂	反相胶束

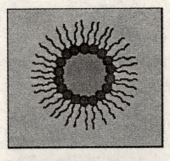 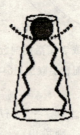

负的或反向曲率
$P>1$
油包水
油溶性胶束

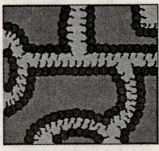 

零或平面曲率
$P\sim1$
双连续

正的或正向曲率
$P<1$
水溶性胶束
水包油型微乳液

图 2.6 表面活性剂分子的临界堆积因子(P_c)变化导致不同的聚集体结构

2.4 液晶中间相

尽管人们对胶束溶液进行了广泛的研究和理论考察,但胶束溶液仅仅是表面活性剂几种可能的聚集态之一。对表面活性剂水溶液行为的全面理解要求有完整的有关表面活性剂所有自组装体系的知识。液晶相的存在是自组装体系中同样重要的一部分,对其详细的描述可参考相关文献(如[44,45]),液晶相的特征总结如下。

2.4.1 定 义

当胶束溶液中表面活性剂的体积分数增加后,尤其是超过起始体积分数约40%时,通常会形成一系列规则的几何结构。由于胶束表面的相互作用为斥力(来自于静电力或水合作用),以至于当聚集体数目增加、胶束间更加靠近时,胶束达到最大分散的惟一办法是改变其形状和大小。这可以解释高浓度下表面活性剂的一系列相行为,这类聚集相被命名为中间相或溶致液晶。

正如其名称所述,液晶兼具有晶体和液体的物理特性:分子排序介于液体和晶体之间,从流变性能来讲,体系既不是简单的粘性液体也不是类似晶体的弹性固体。但液晶相至少在一个方向上是高度有序的,从而在光学上可显示出双折射。

根据体系是表面活性剂还是由其他物质构成,液晶通常被分为两大类,即热致液晶和溶致液晶。热致液晶的结构和性质决定于体系的温度,而溶致液晶的则取决于溶质分子和溶剂分子之间的特殊相互作用。除了天然的脂肪酸皂类表面活性剂,其他表面活性剂液晶都是溶致液晶。

2.4.2 结 构

与表面活性剂-水二元体系有关的主要液晶结构有:六方液晶(正向和反向的),层状液晶和立方液晶(不易确定)。表2.2总结了与这些液晶相有关的一些通用的符号,各液晶相的结构则见图2.6。

表2.2 在表面活性剂-水二元体系中最常见的溶致液晶和其他相态

相结构	符号	其他名称
层状液晶相	L_α	Neat
六方液晶相	H_1	Middle

续表

相结构	符号	其他名称
反式六方液晶相	H_2	
立方液晶相（正向胶束）	I_1	Viscous isotropic
立方液晶相（反向胶束）	I_2	
立方液晶相（正向双连续结构）	V_1	Viscous isotropic
立方液晶相（反向双连续结构）	V_2	
正向胶束	L_1	
反向胶束	L_2	

- 六方液晶是由一系列长圆柱形的棒状胶束（无限长）以六角形紧密堆积排列而成的，这些胶束可能是"正向"的（如水中的 H_1 相），其亲水基在圆柱形棒状胶束外面；或者是"反向"的（H_2），其亲水基在该圆柱形棒状胶束的里面。由于相邻棒状胶束之间的空间充满了憎水基团，反向胶束比 H_1 相中的正向胶束堆积的更紧密，因此 H_2 相在相图中的范围更小且不常见。
- 层状相（L_α）是由交替的水-表面活性剂双分子膜组成的。其疏水链具有极大的自由度和流动性，并且其双分子层结构可以从刚性不可弯曲到可波动弯曲的范围存在。由体系不同，层状液晶的无序程度可以逐渐改变或突然改变，因此，一种表面活性剂可能逐步形成几种截然不同的层状液晶相。
- 立方液晶相的结构变化是多种多样的，并且其在相图的不同部分都可能存在。立方液晶相是典型的光学各向同性体系，所以无法简单地通过偏光显微镜来鉴别。迄今为止，明确定义了两大类立方液晶相：

i. 胶束立方相（I_1 和 I_2）是由小的胶束（或者如 I_2 相中的反相胶束）规则堆积而成。如图 2.7 所示，胶束的形状短而扁长，其排列方式为体心立方形式紧密排列[46,47]。

ii. 双连续立方相（V_1 和 V_2）被认为是在三维空间中能延展很广、多孔、且相互连接的结构。它的形成被认为是由双层结构以及类似分支胶束的互相连接的棒状胶束构成。V_1 相和 V_2 相表明：双连续立方相可以呈正向或反向结构，并且其位置在 H_1 和 L_α 相或 L_α 和 H_2 之间。

除了结构上的差异，液晶相一般在粘度上也有差异，次序如下：

立方液晶相 > 六方液晶相 > 层状液晶相

立方液晶相没有明显的横切平面，表面活性剂的聚集体不易发生相对滑动，

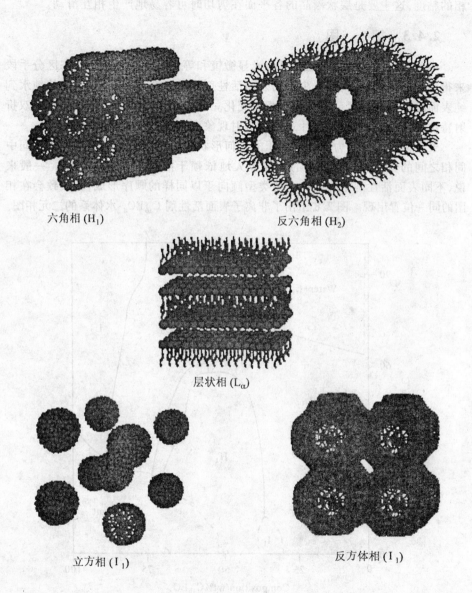

六角相(H_1)　　　　　　　　反六角相(H_2)

层状相(L_α)

立方相(I_1)　　　　　　　　反方体相(I_1)

图2.7　常见表面活性剂液晶相(参见表2.2辨认)

这是立方液晶相一般具有较高的粘度的原因。由于构成六方液晶的圆柱状聚集体只能沿其长轴方向上进行一维移动,六方液晶相虽然一般含有占总体质量30%~60%的水,但却具有较高的粘度。层状液晶相的粘度一般低于六方液晶相的粘度,这主要是层状液晶的各平面在剪切时可容易地产生相互滑动。

2.4.3 相 图

中间相顺序可以简单地通过偏光显微镜和等温技术等常用的相态区分手段来得到。简而言之,就是从少量的表面活性剂开始逐渐增加其浓度,在从纯水到纯表面活性剂的浓度范围内确定相态变化。由于水合晶体和一些液晶具有双折射特性,所以可以通过偏振片观察双折射现象而得到完整的中间相顺序。

聚集体间的作用力和分子堆积的几何形状这两者之间的平衡控制了不同中间相之间的转换。致使体系的特性极大地依赖于溶剂的性质和数量。一般来说,不同表面活性剂中间相的主要类型倾向于以同样的顺序形成,且大致会在相图的同一位置出现。图2.8给出了非离子表面活性剂 $C_{16}EO_8$-水体系的二元相图。

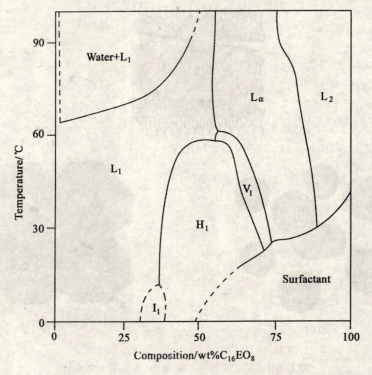

图 2.8 非离子表面活性剂 $C_{16}EO_8$-水体系的二元相图显示了了各种液晶相
(L_1 和 L_2 为各向同性的溶液。其他相态参见表2.2)

(Mitchell et al. *J. Chem. Soc. Faraday Trans. I* **1983**, 79, 975. *Reproduced by permission of the Royal Society of Chemistry*).

参考文献(略)

附录1——表面张力测量方法

应用 Gibbs 方程(见 2.1.2 中的 (2.1.13)式和(2.1.15)式)可以间接得到表面超量。所以在确定体系表面超量方面测量表面张力是一种简便易行的方法。下面详细介绍滴体积法和 du Noüy 环法测量表面张力。

大多数测量平衡界面张力的技术都涉及在测量过程中液-气界面的拉伸。平衡界面张力可以通过应力、压力、液滴尺寸的测量而得到。环法和板法均用于测量应力,而毛细管高度法、最大泡压法则用于测量压力。悬滴法、躺滴法、滴体积法、滴重法以及旋滴法均为测量液滴一维或多维尺寸的方法。

A.1 Du NOüY 环法测量表面张力

环法[A1~A4]:如图 A.1 所示,将铂-铱合金环垂直悬挂在液面上水平浸入液相。

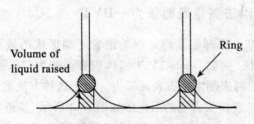

图 A.1 Du Noüy 环示意图

通过计算将环拉出液面所需拉力得到表面张力。假设环上附着液体呈圆柱形,表面张力可从下式得到

$$\gamma_{eq} = \frac{F}{4\pi R} \tag{A.1}$$

式中,R 是环的半径。当平衡时,最大应力为

$$F = (\rho_1 - \rho_2)gV \tag{A.2}$$

式中,ρ_1 和 ρ_2 分别为液相和该液相上方的液相或气相的密度,g 是重力加速度 $(9.81 \text{m} \cdot \text{s}^{-2})$,$V$ 是由环所拉起的液体体积。对于稀水溶液-气相界面,ρ_1 为水密度,ρ_2 为空气密度,所以通过测量由环所带起超出表面的液体重量,就可以得出表面张力。

环法的主要缺点是需要引入一个修正因子。这主要是由环所带起的液体被近似为圆柱形所带来的偏差。根据 Harkins 和 Jordan[A1]的修正因子,(A.1)式可改成

$$\gamma_{eq} = \gamma_{eq}^* \cdot f = \frac{F}{4\pi R} \cdot f \tag{A.3}$$

式中,f 是 Harkins-Jordan 因子,无量纲,γ_{eq}^* 是测量值($mN \cdot m^{-1}$)。

Zuidema 和 Waters 在 Harkins 和 Jordan 修正因子基础上进行进一步推导,得出了计算修正因子的公式(A.4):

$$f = 0.725 + \sqrt{\frac{0.014\,52 \cdot \gamma_{eq}^*}{\frac{1}{4}U^2(\rho_1 - \rho_2)} + 0.045\,34 - \frac{1.679}{R/r}} \tag{A.4}$$

式中,R 是环的平均半径(一般为 10mm),r 是铂-铱合金环的横截面半径(一般为 0.2mm),U 是润湿长度(一般为 120mm)。

参考水在 20℃ 的表面张力可得到用于矫正的最终修正因子,引入已知的环尺寸,并假定气-水界面的密度差$(\rho_1 - \rho_2) = 1$后,得出

$$fk = 1.07\left(0.725 + \sqrt{4.036 \times 10^{-4} \cdot \gamma_{eq}^* + 1.28 \times 10^{-2}}\right) \tag{A.5}$$

A.2 滴体积法测量表面张力—DVT

DVT 方法的基本原理就是测定在毛细管尾端形成的液滴最大尺寸。当今的 DVT 表面张力仪(例如 Lauda TVT1 滴体积表面张力仪)是全自动的,同时可以采用精确体积控制来测定动态表面张力。对于这种方法的详细描述见参考文献[A2,A3]。DVT 表面张力仪的示意图如图 A.2 所示,步进马达推动活塞逐渐降低,使液体逐渐从毛细管顶端流出,活塞持续推动液滴增大,直到液滴重量(mg)超过表面张力上拉的力($2\pi r_{cap}\gamma$)后停止。此时仪器采用光检测器测量液滴脱离毛细管的下落运动。通过液滴的体积 V 并由(A.6)式[A.4]即得出表面张力 γ 为

$$\gamma = \frac{V\Delta\rho g}{2\pi r_{cap}}f \tag{A.6}$$

式中,$\Delta\rho$ 是两相密度差,g 是重力加速度,r_{cap} 是毛细管半径;f 是修正因子,这个修正因子主要是为了修正液滴并不是在毛细管顶端脱离,而是在液滴本身的颈部脱离而引入的[A.5]。

A.3 活度系数的计算

当研究离子表面活性剂的表面张力随浓度变化时,常常用到活度而不是浓

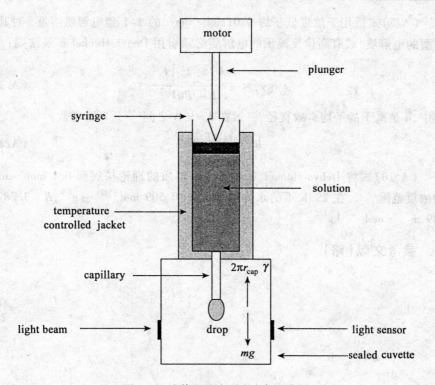

图 A.2　滴体积法表面张力仪示意图

度。在极稀的溶液条件下,比如低于 1×10^{-3} mol·dm^{-3} 浓度时,活度系数可以被认为等于 1,但在大于 1×10^{-3} mol·dm^{-3} 的浓度下,活度系数不能被认为等于 1,此时离子间库仑力的相互作用使体系偏离了理想状态,需要引入有关离子强度的 Debye-Hückel 理论。文献[A.6,A.7]对此进行了详细介绍,这里仅给出有关的公式。在较低电解质浓度下,平均活度系数 $\gamma_\pm$ 可以由 Debye-Hückel 公式计算:

$$\lg\gamma_\pm = -a|z_+ z_-|I^{1/2} \tag{A.7}$$

式中,z 为电解质离子电荷数,I 是离子强度,A 是常数。下面给出了 I 和常数 A 的表达式:

$$I = \frac{1}{2}\sum_i m_i z_i^2 \tag{A.8}$$

$$A = \frac{F^3}{4\pi N_a \ln 10}\left(\frac{\rho}{2(\varepsilon_o \varepsilon_r RT)^3}\right)^{1/2} \tag{A.9}$$

式中,m 是重量摩尔浓度,z 是离子价数,ρ 是溶剂密度。F,N_a,R,ε_o 和 ε_r 是标准物理常数。

（A.7）式适用于浓度低于约 $0.01 \text{mol} \cdot \text{dm}^{-3}$ 的 1∶1 型电解质溶液。对其他类型的电解质，或有高价反离子的电解质必须引用 Debye-Hückel 扩展法则：

$$\lg \gamma_{\pm} = -\frac{A|z_+ z_-|I^{1/2}}{1 + BaI^{1/2}} \quad (A.10)$$

式中，a 是离子的平均有效直径，一般在 3～9 Å 之间[A8]。B 是常数：

$$B = \left(\frac{2F^2\rho}{\varepsilon_o\varepsilon_r RT}\right)^{1/2} \quad (A.11)$$

（A.10）式将 Debye-Hückel 对 1∶1 型电解质的理论扩展到 $0.1 \text{ mol} \cdot \text{dm}^{-3}$ 的浓度范围[A7]。在 298K 下的水溶液中，$A = 0.509 \text{ mol}^{-1/2} \cdot \text{kg}^{1/2}$，$B = 3.282 \times 10^9 \text{ m}^{-1} \cdot \text{mol}^{-1/2} \text{ kg}^{1/2}$。

参考文献（略）

第三章 微 乳 液

本章主要讨论表面活性剂的另一重要性质,即水-油界面的稳定性和微乳液的形成。由于能够溶解在其他溶剂中不溶的物质(如水相中的非极性化合物),这种特殊的胶体分散体系引起了广泛的关注。在过去的四十年中,微乳的工业应用进一步扩大,对微乳的形成、稳定性及表面活性剂分子构造的原理的理解也有了进一步的提高。本章介绍与目前工作相关的主要理论和用于表征微乳的常用技术。

3.1 微乳液的定义和历史

微乳液的最好定义之一是 Danielsson 和 Lindman[1]共同给出的:"微乳液是水、油和两亲分子组成的各向同性的、热力学稳定的溶液体系。"在某些方面,微乳液被认为是小液滴的乳液。例如,液滴半径在 5~50nm 之间的分散类型为 O/W 和 W/O 的小液滴。由于微乳液和通常的乳液(粗乳状液)有很重要的区别,所以这样的说法欠推敲。特别是当乳液中粒径的平均尺寸不断地随时间增加,最终的相分离是在重力作用下进行的,也就是说乳液是热力学不稳定的体系,乳液的形成需要外界做功。乳液的分散相液滴一般较大($>0.1\ \mu m$),因此外观为乳状,而非透明。而微乳液,在一定的条件下会自发形成。对于简单的含水体系,微乳液的形成依赖于表面活性剂的类型和结构。如果表面活性剂是离子型的且只有单一烃链(如十二烷基磺酸钠,SDS),若在助表面活性剂(如中等大小的脂肪醇)和/或电解质(如 0.2 M NaCl)存在下就可形成微乳液。如果是双链的离子型表面活性剂(如 Aerosol-OT)和一些非离子型表面活性剂,助表面活性剂就不是必需的。这是微乳液最基本的性质之一,即油相和水相之间的超低界面张力 $\gamma_{o/w}$ 作用的结果:表面活性剂的主要作用就是将 $\gamma_{o/w}$ 降到足够低(也就是说降低能量需要增加表面积),以便水滴和油滴发生自发分散且体系是热力学稳定的。在 3.2.1 中将要论述到超低界面张力的形成是微乳液形成的关键,它与体系的组成有关。

直到 1943 年 Hoar 和 Schulman 报道了当强表面活性剂加入到水和油中自

发形成乳液[2]后人们才真正认识了微乳液。"微乳"一词的使用,最初是 Schulman 等[3]在 1959 年描述由水、油、表面活性剂和醇形成的一个透明溶液体系时提到的。对于用"微乳"来描述这样的体系存在许多的争论[4]。现在,由于还没有系统地使用,人们更愿意称之为"胶状乳液"[5]或"胀大的胶束"[6]。微乳液可能在 Schulman 研究之前就被发现,20 世纪澳大利亚家庭主妇就用水/桉树油/皂片/松节油混合物洗羊毛,第一种商品微乳液也许是 1928 年 Rodawald 发现的液体石蜡。在 20 世纪 70 年代末、80 年代初微乳的应用有了更进一步的发展,当时,发现微乳体系能够提高采油量,在当时的油价水平上,利用三次采油可获取更多的利润[7]。现在,情况不再是这样了,但微乳在其他方面的应用也为人所知,如催化作用、纳米粒子的制备、太阳能转化、液液萃取(矿物、蛋白质等)。加上洗涤和润滑的经典应用,这个领域仍然有足够的重要性来吸引许多的科学家。从基本研究的观点来看,很多的进步是在了解微乳液性质后的最近 20 年中取得的。尤其是新的、有效技术的发展,如小角中子散射(SANS),使解释界面膜稳定性和微乳结构成为可能。下面主要介绍微乳液的基本性质,即微乳的形成和稳定性、表面活性剂膜、分类和相行为。

3.2 形成和稳定性理论

3.2.1 微乳的界面张力

描述微乳液形成的一个简单形式是把分散相考虑成很小的液滴,其构型熵变化(ΔS_{conf})可近似地表示为[8]

$$\Delta S_{conf} = - nk_B [\ln\phi + \{(1-\phi)/\phi\}\ln(1-\phi)] \qquad (3.2.1)$$

式中,n 为分散相的液滴数,k_B 为 Boltzmann 常数,ϕ 是分散相的体积分数。缔合自由能的改变可用新增加的界面面积所需的自由能 $\Delta A\gamma_{12}$ 和构型熵之和来表示[9]:

$$\Delta G_{form} = \Delta A\gamma_{12} - T\Delta S_{conf} \qquad (3.2.2)$$

式中,ΔA 是界面面积 A 的改变量(半径为 r 的液滴面积为 $4\pi r^2$),γ_{12} 是在温度 T(Kelvin)的 1 相和 2 相(如油相和水相)之间界面张力。用(3.2.2)式取代(3.2.1)式,可求出 1 相和 2 相之间的最大界面张力。分散的小液滴数目增加使得 ΔS_{conf} 为正值,如果表面活性剂能将界面张力降到足够低,(3.2.2)式中的能量项 $\Delta A\gamma_{12}$ 相对较小,这样得到有利的负的自由能变化,就可自发形成微乳液。

在无表面活性剂的油-水体系中,$\gamma_{o/w}$ 在 50 mN·m^{-1} 左右,在形成微乳液的过程中,界面面积 ΔA 较大,通常量级是 $10^4 \sim 10^5$。这样,在无表面活性剂时(3.2.2)式中的第二项数量级就是 $1\,000\,k_B T$,为了满足 $\Delta A\gamma_{12} \leqslant T\Delta S_{conf}$ 的条件,界面张力应该非常低(约 $0.01\,\text{mN}\cdot\text{m}^{-1}$)。一些表面活性剂(双链离子型[10,11]和一些非离子型的[12])能将界面张力降到很低($10^{-2} \sim 10^{-4}\,\text{mN}\cdot\text{m}^{-1}$),但在大多数情况下,由于在获得低界面张力之前就已达到表面活性剂的 CMC,这样低的界面张力并不能由单一表面活性剂来获得。要进一步降低界面张力的有效方法就是加入第二种具有表面活性的物质即助表面活性剂(如一种表面活性剂或中等链长的醇)。这可从 Gibbs 方程扩展为多组分体系后得到理解[13]。界面张力与体系中表面活性剂膜的组成和每一组分的化学势 μ 有关,即

$$d\gamma_{o/w} = -\sum_i (\Gamma_i d\mu_i) \approx -\sum_i (\Gamma_i RT d\ln C_i) \tag{3.2.3}$$

式中,C_i 是 i 组分在混合物中的摩尔浓度,Γ_i 是表面过剩($\text{mol}\cdot\text{m}^{-2}$)。假定浓度分别为 C_s 和 C_{co} 的表面活性剂和共表面活性剂是仅有的吸附组分(即 $\Gamma_{\text{water}} = \Gamma_{\text{oil}} = 0$),则(3.2.3)式就变为

$$d\gamma_{o/w} = -\Gamma_s RT d\ln C_s - \Gamma_{co} RT d\ln C_{co} \tag{3.2.4}$$

综合(3.2.4)式可得

$$\gamma_{o/w} = \gamma^{\circ}_{o/w} - \int_0^{C_s} \Gamma_s RT d\ln C_s - \int_0^{C_{co}} \Gamma_{co} RT d\ln C_{co} \tag{3.2.5}$$

(3.2.5)式说明 $\gamma^{\circ}_{o/w}$ 是由表面活性剂和助表面活性剂的表面过剩两相来降低的(表面过剩分别为 Γ_s 和 Γ_{co}),因此它们的效果是叠加的。然而,值得注意的是,这两个分子应是同时吸附且彼此不发生相互作用(否则它们就会降低彼此的活性),也就是说,这两种分子应具有完全不同的化学性质,不能形成混合胶束。对于特定的表面活性剂和助表面活性剂体系,界面张力可以降至很低,以至于进一步增加浓度也不会使 $\gamma_{o/w}$ 成为负值。Overbeek[14]研究了盐水-环己烷-正戊醇-SDS 体系的 $\gamma_{o/w}$ 随浓度的变化(见图 3.1):表面活性剂和共表面活性剂过量时界面的自发膨胀会使 $\gamma_{o/w}$ 产生短暂的负值。图 3.1 中 20% 戊醇的图有进一步的说明。然后,$\gamma_{o/w}$ 又达到一小的正平衡值,自发形成微乳。这样的热力学处理最初是由 Ruckenstein 和 Chi[15]和 Overbeek[8]提出的。(3.2.1)式和(3.2.2)式是非常近似的;尤其是所添加的项可用于说明液滴之间的相互作用[15]。

3.2.2 动力学稳定性

人们知道微乳液滴的内部组分可发生交换,典型的是在毫秒级的范围[16,17]。液滴扩散并相互碰撞,如果碰撞十分剧烈,表面活性剂膜就会发生破裂并促使液滴间的交换,这就是液滴的动力学不稳定性。然而,对于一个非常小

液滴(<500Å)的体系,液滴合并要克服的能垒较大,这个体系将会在较长一段时间(几个月)内保持分散、透明的状态[18]。这样的微乳被认为是动力学稳定的[19]。液滴合并的机理在 AOT 的 W/O 型微乳液中已有报道[16];液滴交换过程是以二阶速度常数 k_{ex} 为特征的,这个常数不仅由扩散控制,还受活化作用控制(即提高活化能 E_a 对溶合产生阻碍)。其他研究[20]也表明:微乳液的动力学特征体现在界面的柔性,即膜的刚性(见 3.3.3 节)对能垒的影响。在同一实验条件下,不同的微乳体系有不同的 k_{ex} 值[16]:室温下 AOT W/O 体系的 k_{ex} 在 $10^6 \sim 10^9$ $dm^3 \cdot mol^{-1} \cdot s^{-1}$ 范围内,非离子 C_iE_j 的 k_{ex} 在 $10^8 \sim 10^9$ $dm^3 \cdot mol^{-1} \cdot s^{-1}$ 范围[16,17,20]。无论如何,总能够保持一个平衡小液滴的外形和大小,这样就可用不同的方法进行研究[20]。

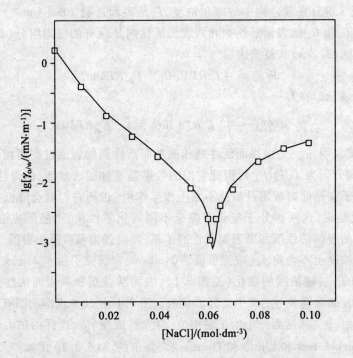

图 3.1 在 AOT 存在下正庚烷-NaCl 水溶液中的油-水界面张力与盐浓度的函数关系(界面张力值用旋滴张力仪测定,AOT 浓度为 0.050 $mol \cdot dm^{-3}$,温度为 25℃)

3.3 物理化学性质

本节介绍表征微乳的主要参数。相关的参考文献已在第二章介绍。

3.3.1 微乳液类型的预测

众所周知的微乳液分类是由 Winsor 提出的[21]，他命名了四种相平衡类型。

- 类型 Ⅰ：水溶解性好的表面活性剂形成 O/W 型微乳液（Winsor Ⅰ 型）。富含表面活性剂的水相与油相共存，油相中表面活性剂仅以浓度较低的单体形式存在。
- 类型 Ⅱ：表面活性剂主要存在于油相中形成 W/O 型微乳液。富含表面活性剂的油相与含表面活性剂较少的水相共存（Winsor Ⅱ 型）。
- 类型 Ⅲ：是三相体系，富含表面活性剂的中间相与含较少表面活性剂的水、油两相共存（Winsor Ⅲ 型或中相微乳）。
- 类型 Ⅳ：单相（各向同性）的胶束溶液，加入足够的两亲分子（表面活性剂加醇）时形成。

Winsor Ⅰ、Ⅱ、Ⅲ 或 Ⅳ 型的形成与表面活性剂的类型和样品环境有关，占主导地位的类型与表面活性剂分子在界面上的分布有关（见下面部分）。图 3.2 中示出的是由电解质浓度的增加（对于离子型表面活性剂）或温度的升高（对于非离子型表面活性剂）而产生的相转变。表 3.1 总结了当组成改变时阴离子表面活性剂相行为定性的变化[22]。

许多研究者致力于吸附界面膜相互作用的研究用以解释界面曲率的方向和大小。第一个概念由 Bancroft[23] 和 Clowes[24] 提出，他们认为在乳液体系中的界面吸附膜是双层的，内层和外层的界面张力独立起作用[25]。对于界面弯曲，内层的表面张力较高。Bancroft 的理论是基于"乳化剂最易溶的相为外层相"，即，油溶性的表面活性剂形成 W/O 型微乳液，水溶性的表面活性剂形成 O/W 型微乳液。这一定性的概念在很大程度上被进一步扩充，一些参数已被建议来定量描述表面活性剂膜的性质。关于三种微乳的类型和在相图中的位置的进一步细节，将在 3.3.3 中介绍。

1. R 比理论

Winsor 首先用 R 比理论说明两亲分子和溶剂对界面曲率的影响[21]。它最初用来描述两亲分子单分子层和油、水区域之间的相互作用的能量。因此，R 比从能量角度将两亲分子分散于油中的趋势与溶解于水中的趋势进行了比较。如果有利于其中一相,界面区域就呈现一定的曲率。下面简单描述这个概念,在其他文献还中有更详尽的说明[26]。

胶束和微乳液可被划分为三种截然不同(单相和多相)的区域:水相(W)、油相或有机相(O)、两相区(C)。如图 3.2 所示,将界面区当做是含有确定组分且水相与油相是完全分开的,其中界面区的厚度是有限的,除了表面活性剂分子外,还含有一些油和水。

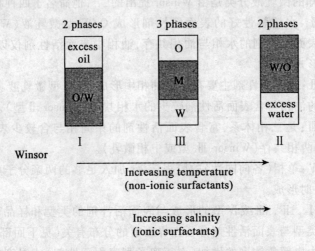

图 3.2 分别对非离子型和离子型表面活性剂进行温度和盐度扫描的 Winsor 分类和相序结果(大部分表面活性剂处于阴影区域中。在三相体系中中相微乳液(M)与过剩的水相(W)和油相(O)处于平衡)

表 3.1 非离子表面活性剂溶液中可观察到的相行为与定性因素的关系

扫描时的变量(增加)	三相图的转变
盐度	Ⅰ ∈ Ⅲ ∈ Ⅱ
油:烷基碳链数	Ⅱ ∈ Ⅲ ∈ Ⅰ
乙醇:低分子量	Ⅰ ∈ Ⅲ ∈ Ⅱ
高分子量	Ⅱ ∈ Ⅲ ∈ Ⅰ
表面活性剂:亲油性链长	Ⅰ ∈ Ⅲ ∈ Ⅱ
温度	Ⅱ ∈ Ⅲ ∈ Ⅰ

(见 Bellocq et al.[22])

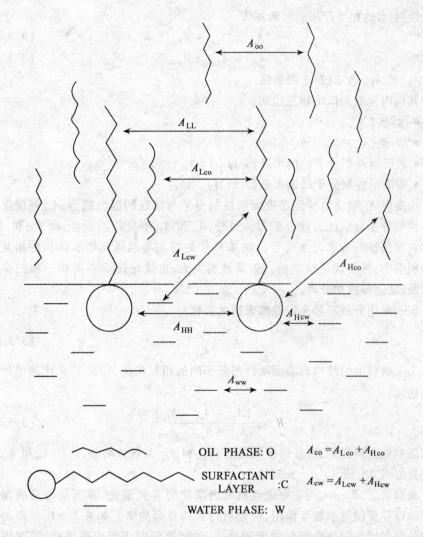

图 3.3 油-表面活性剂-水体系界面区的相互作用能

内聚能存在于 C 层内且决定界面膜的稳定性。简要的示意见图 3.3，分子 X 和 Y 之间的内聚能定义为 A_{xy}，分子间发生相互作用时是正值。A_{xy} 是在各向异性的界面 C 中表面活性剂与油、水在单位面积上作用的内聚能。对于表面活性剂-油和表面活性剂-水的相互作用，A_{xy} 包括两部分：

$$A_{xy} = A_{Lxy} + A_{Hxy} \tag{3.3.1}$$

式中，A_{Lxy} 是两个分子非极性部分的定量作用（典型的 London 分散力），A_{Hxy} 是极性部分的定量作用，尤其是氢键键合和库仑力作用。这样，表面活性剂 - 油和表

面活性剂-水的相互作用就可表示为

$$A_{co} = A_{Lco} + A_{Hco} \tag{3.3.2}$$

$$A_{cw} = A_{Lcw} + A_{Hcw} \tag{3.3.3}$$

A_{Hco} 和 A_{Lcw} 通常较小,可忽略。

其他内聚能为以下相互作用:
- 水-水,A_{ww};
- 油-油,A_{oo};
- 表面活性剂分子的憎水基部分(L),A_{LL};
- 表面活性剂分子的亲水基部分(H),A_{HH}。

内聚能 A_{co} 明显地可促进表面活性剂分子与油区的混合能力,A_{cw} 可促进表面活性剂分子与水区的混合能力。另外,A_{oo} 和 A_{LL} 不易与油混合,而 A_{ww} 和 A_{HH} 阻碍表面活性剂分子与水混合。如果 C 中的溶剂与油体相和水体相的相互作用差别很小,界面就是稳定的。如果差别太大,也就是说 C 与其中一相的亲和力太强,就会导致相分离。

Winsor 用分散趋势来定性的表征此变化:

$$R = \frac{A_{co}}{A_{cw}} \tag{3.3.4}$$

为了解释油的结构和表面活性剂分子间的相互作用,提出了 R 比理论的最初表达式:

$$R = \frac{(A_{co} - A_{oo} - A_{LL})}{(A_{cw} - A_{ww} - A_{HH})} \tag{3.3.5}$$

正如前面所提到的,在许多情况下,A_{Hco} 和 A_{Lcw} 是被忽略的,这样 A_{co} 和 A_{cw} 分别被近似为 A_{Lco} 和 A_{Hcw}。

简而言之,Winsor 的主要观点就是内聚能的 R 比理论,界面层与油的相互作用除以界面层与水相互作用,决定所得到的界面曲率。如果 $R>1$,界面与油接触的面积增加而与水接触的面积减少,油相就倾向于形成连续相,即对应的 Winsor Ⅱ型。类似的,平衡的界面层 $R=1$。

2. 堆积系数和微乳液结构

界面曲率和微乳液类型可以用几何学来进行定量的表示。这个概念是由 Israelachivili 等提出[27],并被广泛应用于与界面分布有关的表面活性剂分子结构。正如在 2.3.3 中所述,界面曲率是由亲水基相对面积 a_o 和亲油基的相对面积 v/l_c 决定的(见图 2.6)。就微乳液类型来说:
- 如果 $a_o > v/l_c$,形成 O/W 型微乳液;
- 如果 $a_o < v/l_c$,形成 W/O 型微乳液。

- 如果 $a_o = v/l_c$,形成中相微乳液

3. HLB 理论

另一个与分子结构、界面堆积、膜曲率有关的概念是亲水亲油平衡(HLB)。它通常被表示为基于分子中亲油基团和亲水基团相对比例的经验方程。这个概念最初是由 Griffin 提出[28],他表征了许多表面活性剂,并基于表面活性剂的化学组成提出了针对非离子的烷基聚乙二醇乙醚(C_iE_j)的经验方程。

$$HLB = (E_j wt\% + OH wt\%)/5 \qquad (3.3.6)$$

式中,$E_j wt\%$ 和 $OH wt\%$ 分别表乙氧基和羟基的重量百分数。

Davies 等[30]提出了一个更普通的经验公式,表达了不同的亲水亲油基团与 HLB 的关系:

$$HLB = [(n_H \times H) - (n_L \times L)] + 7 \qquad (3.3.7)$$

式中,H 和 L 分别为亲水、亲油基团的常数,n_H 和 n_L 分别为每个表面活性剂分子中这些基团的个数。

对于双连续结构,也就是曲率为零,$HLB \approx 10$[31]。当 $HLB < 10$ 时,形成 W/O 型微乳液,$HLB > 10$ 时形成 O/W 型微乳液。HLB 和堆积系数描述了相同的基本概念,而堆积系数更适合于微乳液。表面活性剂几何学、体系的 HLB 值和堆积系数的影响在图 3.4 中进行了说明。

4. 相转变温度(PIT)

油包水型微乳液(乳状液)中非离子表面活性剂对温度的敏感性较高,特别是它在相转变温度(PIT)时膜曲率由正变为负。Shinoda 等对此做了规定[32]:

- 如果 $T < PIT$,形成水包油型微乳液(Winsor Ⅰ);
- 如果 $T > PIT$,形成油包水型微乳液(Winsor Ⅱ);
- 如果 $T = PIT$,存在一个中相微乳液(Winsor Ⅲ),而且其自发曲率为零,HLB 值大约为 10。

HLB 值和 PIT 是相互联系的,因而有时也使用 HLB 温度。

3.3.2 表面活性剂膜的性质

油水界面的表面活性剂膜的物理特性可用三个物理参数来表述:张力、弯曲刚性和自发曲率。它们的相对重要性取决于膜受到限制的大小。由于表面活性剂膜决定微乳液(乳状液)的静态和动态稳定性,理解这些参数与界面稳定性之间的关系是非常重要的,包括相行为、稳定性、结构和增溶能力。

1. 超低的界面张力

第二章定义了平面界面(表面)张力 γ,同样的原则可用于弯曲的液-液界面,也就是说,它适用于所有增加单位界面面积所作功。如 3.2.1 所述,微乳液

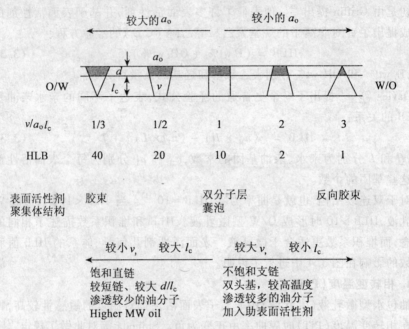

图 3.4 分子几何构型和体系条件对包裹参数和 HLB 值的影响
(参见 Israelachvili[31])

形成时须具有超低的界面张力,$\gamma_{o/w}$,一般为 $10^{-2} \sim 10^{-4}$ mN·m^{-1}。界面张力受助表面活性剂、电解质、温度、压力和亲油链长度的影响。影响 $\gamma_{o/w}$ 的一些变量的作用有很多研究报道,值得一提的是,Aveyard 及其合作者通过改变亲油链长度、温度和电解质浓度对离子型[34,35]和非离子型[36]表面活性剂的表面张力进行了系统研究。例如,在水—AOT—正庚烷体系中,当表面活性剂浓度大于 CMC 一定量时,在 $\gamma_{o/w}$ 与电解质(NaCl)浓度的函数图中,在 Winsor 相转变处出现最小值,也就是说,随着 NaCl 浓度的增加,$\gamma_{o/w}$ 逐渐降低至最小值(Winsor Ⅲ 构型),然后又增加至大约 $0.2 \sim 0.3$ mN·m^{-1}(Winsor Ⅱ 区域)。在电解质浓度一定的条件下,改变温度[34]、亲油链长度和助表面活性剂的浓度[35]有类似的结果。对于非离子型表面活性剂,随着温度的升高,可以看到类似的张力曲线和相转变[36]。另外,随着表面活性剂亲油链长的增加,最小界面张力出现在较高的温度,且变小[37]。对于超低界面张力,一些标准技术手段,如 Du Nouy 环法,Wilhelmy 板法或滴体积法(DVT)是检测不到的,合适的检测技术有旋滴法(SDT)和表面光散射法[38]。

2. 自发曲率

自发曲率(自然或起始曲率)C_o 是在等量的水和油体系中所形成的表面活性剂膜的曲率。膜上没有任何外加力,处于最低的自由能态。但当其中一相占有优势时,C_o 会有一定的偏离。一般来说,界面上的每一个点都有两个基本的曲率半径,R_1 和 R_2,它们与基本曲率之间的关系为 $C_1 = 1/R_1$ 和 $C_2 = 1/R_2$。一般用平均曲率和高斯曲率来表示界面弯曲度,它们的定义为[39]

$$平均曲率:C = 1/2(1/R_1 + 1/R_2)$$

$$高斯曲率:\kappa = 1/R_1 \times 1/R_2$$

C_1 和 C_2 是这样确定的:表面活性剂膜的表面上的每一个点都有两个基本曲率半径 R_1 和 R_2,如图 3.5 所示。假设放置一个圆环在表面上的某个切点上,如果这个圆的半径恰好使其在接触点上的二次导数与表面上在这个方向上的切线(向量,n)的二次导数相同,那么这个圆环的半径就是表面的曲率半径。表面曲率由这样的两个相垂直的圆来表述,如图 3.5(a)所示。

对于球体,R_1 和 R_2 相等并且同为正数(见图 3.5(b));对于圆柱体,R_2 为无穷大(见图 3.5(c));对于平面,R_1 和 R_2 均为无穷大;对于特殊的马鞍形,$R_1 = -R_2$,也就是说,界面上的每一点均有两个既凸又凹具有相同曲率半径的圆(见图 3.5(d))。平面和马鞍形的平均曲率均为零。

曲率 C_o 由分离时两相的组成和表面活性剂的类型所决定。一种应用在界面的非极性一侧的观点认为:油在一定程度能渗透至表面活性剂的烃基链之间。渗透范围越大,曲率越向极性的一面弯曲。一般地,正曲率朝向油(负曲率朝向水),因此会引起 C_o 的下降。亲油链越长,越难以渗透表面活性剂膜,对 C_o 的影响也越小。最近,Eastoe 等用 SANS 选择性氘代烃链研究了双链表面活性剂稳定的微乳液的溶剂渗透程度,结果表明,油渗透是一相当敏感的效应,主要取决于油和表面活性剂的化学结构。另外,表面活性剂链长的不同和不饱和键的存在使表面活性剂/油的界面变得更加无序,油混合程度更高。然而,对于对称的双链表面活性剂(例如 DDAB 和 AOT)没有发现油混合的证据[42]。采用庚烷和环己烷研究了烷基结构和分子体积对油渗透的影响,结果发现,正庚烷不在界面层上,而结构紧密的环己烷有更明显的渗透效果[43]。

表面活性剂类型和亲水基的性质也通过极性相(水相)间的不同相互作用而影响 C_o。

• 对于离子型表面活性剂,电解质浓度和温度对自发曲率的影响相反,盐浓

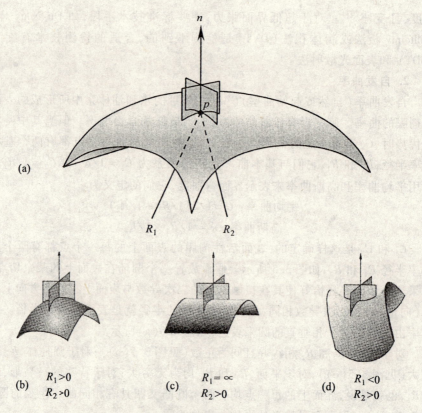

(a) 含普通向量(n)的平面上的点 P 对表面活性剂膜平面的分割；
(b) 凸曲率；(c) 圆柱体曲率；(d) 鞍形曲率
图 3.5　表面上不同的曲率原则(Hyde et al.[39])

度的增加屏蔽了带静电的亲水基间的斥力，即减小了亲水基的面积，引起表面活性剂膜向水相弯曲，因此使 C_o 变小。温度的升高有两方面的影响：(1) 多价的带相反电荷的离子的分裂导致亲水基静电斥力的增加，使 C_o 变大；(2) 表面活性剂支链越多、越卷曲，C_o 越小。因此温度对非极性链和静电相互作用的影响是互相竞争的。一般认为静电效应占有一定优势，所以 C_o 随温度的增加略有增加。

- 对于非离子型表面活性剂，电解质对 C_o 的影响很小，而温度对非离子型表面活性剂溶解度的影响较大（水或油中），因此是一个重要的参数。对于 C_iE_j 型表面活性剂，随着温度升高，亲水基在水中的溶解度变小并且水也更难渗透至表面活性剂层；另外，在膜的另一面，油更易渗透至烃链中,因此,对于这种非离子型表面活性剂,温度增加引起 C_o 的强烈降低。这个现象解释了图2.8(见第二章)中温度对相平衡的强烈影响。

因而,通过改变外部参数如温度、油的性质或电解质浓度,自发曲率可被调节到适当的值,而且可促进 Winsor 体系间的相互转化。另外还有一些以相似的方式影响 C_o 的因素,包括亲水基的改变、带有相反电荷离子的类型和价态、非极性链的长度和数量、助表面活性剂或混合表面活性剂的加入。

3. 膜弯曲刚性

膜刚性是与界面曲率相联系的一个重要参数。膜弯曲能的概念首先由 Helfrich[45] 提出,现在已被认为是理解微乳液性质的一个基本模型。膜弯曲能可以用两个弹性系数[46]来表述,它们是用来测量将界面膜由预定的平均曲率改变所需的能量:

- 平均弯曲弹性(或刚性)K,与平均曲率相关,表示将单位面积的表面弯曲单位数量所需的能量,K 为正值,有利于自发曲率。
- 弹性 $\bar{K}$,与高斯曲率相关,表示膜的拓扑学,其单位为焦耳,对于球形结构为负值,双连续的立方晶相为正值。

理论上讲,弯曲系数主要与表面活性剂链长[47]、表面活性剂分子在膜上的面积[48]以及带有静电的亲水基间的相互作用有关[49]。

膜刚性理论是基于与膜曲率相关联的界面自由能的。而处于界面的表面活性剂膜的自由能 F,可以由界面能 F_i,弯曲能 F_b 以及熵变 F_{ent} 的和得到[50]。

$$F = F_i + F_b + F_{ent} = \gamma A + \iint \left[\frac{K}{2}(C_1 + C_2 - 2C_o)^2 + \bar{K} C_1 C_2 \right] dA + n k_B T f(\phi) \tag{3.3.8}$$

式中,γ 为界面张力,A 为膜的总面积,K 为平均弹性弯曲系数,$\bar{K}$ 为高斯弯曲系数,C_1 和 C_2 为两个基本曲率,C_o 为自发曲率,n 为液滴数,k_B 为 Boltzmann 常数,$f(\phi)$ 为微乳液滴的混合熵的函数,ϕ 为液滴核部分的体积分数。若 $\phi < 0.1$,$f(\phi) = [\ln(\phi) - 1]$[50]。由于微乳液形成时具有超低的界面张力,因此 A 与 F_b 和 F_{ent} 相比可以忽略不计。

如前所述,曲率 C_1,C_2 和 C_o 可由 $1/R_1$,$1/R_2$ 和 $1/R_o$ 等半径的形式来表示,对于球形液滴,$R_1 = R_2 = R$,界面的表面积为 $A = n 4\pi R^2$。注意 R_1 和 R_o 为核半径而不是液滴半径[50]。解(3.3.8)式并除以 A,对于球形液滴(半径为 R),总自由能 F,可作如下表述:

$$\frac{F}{A} = 2K \left(\frac{1}{R} - \frac{1}{R_o} \right)^2 + \frac{\bar{K}}{R^2} + \left[\frac{k_B T}{4\pi R^2} f(\phi) \right] \tag{3.3.9}$$

对于达到增溶极限的体系（WⅠ或WⅡ区域），微乳液与过剩的溶液相相平衡，并且液滴达到最大值。也就是说，最大的核半径R_{max}^{av}，在总自由能最小的前提下，自然曲率半径R_o与弹性系数K和$\bar{K}$之间的关系如下：

$$\frac{R_{max}^{av}}{R_o} = \frac{2K+\bar{K}}{2K} + \frac{k_BT}{8\pi K}f(\phi) \qquad (3.3.10)$$

有许多技术可用来测量K和$\bar{K}$，特别是椭圆偏光法、X射线反射法和小角度X射线散射法（SAXS）[52-54]。De Gennes和Taupin[55]发展了一个双连续微乳液的模型，对于$C_o=0$，在没有热量波动的平面液层，他们由下式引入ξ_K，它是与K有关的表面活性剂层的保留长度。

$$\xi_K = a\exp(2\pi K/k_BT) \qquad (3.3.11)$$

式中，a为分子长度，ξ_K为膜层正常长度的关联长度，也就是在热波动存在时液层保持平面所需的距离。ξ_K对于K的大小相当敏感，当$K \gg k_BT$时，ξ_K为宏观量，也就是说表面活性剂层大范围为平面，可能形成层状结构。如果降低到$\sim k_BT$，那么ξ_K为微观的，有序结构是不稳定的而且可能形成无序的相，如微乳液。实验表明，对于浓缩的不溶解的单层，K在$100k_BT$内[56]；对于脂质双分子层，K在$10k_BT$内[57-59]，而在微乳液体系中K可降低到k_BT以下[60]。$\bar{K}$的角色也是相当重要的，然而，文献中很少有相关的报道[e.g., 53,61]，在决定表面活性剂-油-水混合结构的重要性时依旧不确切。

另一种更可行的测定膜刚性的方法是计算由张力仪和SANS测量的复合参数$(2K+\bar{K})$，对于处于增溶边界、WⅠ或WⅡ体系中的液滴微乳液，该参数可结合液滴的半径、界面张力和液滴的多分散性而得到。从(3.3.9)式和(3.3.10)式可得到下面两种表达式来描述。

(1) 界面张力$\xi_K\gamma_{o/w}$（由SLS或SDT测得）和平均核半径的最大值R_{max}^{av}（由SANS测得）：

$$2K + \bar{K} = \gamma_{o/w}(R_{max}^{av})^2 - \frac{k_BT}{4\pi}f(\phi) \qquad (3.3.12)$$

在增溶边界上，微乳液和过剩相之间的界面张力$\gamma_{o/w}$可由弹性系数和R_{max}^{av}来表述[52]。既然表面活性剂单分子层是在弯曲的微乳液滴的周围形成，任何新产生的面积都为界面面积，能量就可以在WⅠ或WⅡ体系中计算[56]。为做到这一点，表面活性剂膜就必须具有拉伸性，由此引入$\bar{K}/(R_{max}^{av})^2$。而所产生液滴数量的改变引起熵的改变，熵变分布中包含了$\bar{K}$，$\bar{K}/(R_{max}^{av})^2$是拓扑上引起的改变，

因此微乳液和过剩相之间的界面张力可表达为

$$\gamma_{o/w} = \frac{2K + \bar{K}}{(R_{max}^{av})^2} + \frac{k_B T}{4\pi (R_{max}^{av})^2} f(\phi) \qquad (3.3.13)$$

弯曲系数可由下式给出：

$$2K + \bar{K} = \gamma_{o/w} (R_{max}^{av})^2 - \frac{k_B T}{4\pi} f(\phi)$$

(2)使用由 SANS 分析得到的 Schultz 多分散性宽度 $p = \sigma/R_{max}^{av}$：

液滴多分散性与由热波动引起的微乳液滴的弯曲系数有关。Safran[62]和 Milner[63]依据球谐波描述了由液滴变化引起的热波动,他们发现引起波动的基本原因是波形 $l = 0$ 的变形[50],而引起液滴大小的改变,也就是平均液滴半径的改变并因此而引起液滴分散性的改变。在最大增溶态(WⅠ或WⅡ)当两相平衡时,多分散性 p 是 K 和 $\bar{K}$ 的函数：

$$p^2 = \frac{u_o^2}{4\pi} = \frac{k_B T}{8\pi(2K + \bar{K}) + 2k_B T f(\phi)} \qquad (3.3.14)$$

式中,u_o 表示 $l = 0$ 时的波振幅,多分散性在 SANS 中由 Schultz 多分散参数 R_{max}^{av} 给出[64],(3.3.14)式可写为

$$2K + \bar{K} = \frac{k_B T}{8\pi(\sigma/R_{max}^{av})^2} - \frac{k_B T}{4\pi} f(\phi) \qquad (3.3.15)$$

(3.3.12)式和(3.3.15)式利用 SANS 和张力仪的数据给出了 $(2K + \bar{K})$ 两个相近的表述。这种方法被用来描述非离子型表面活性剂形成的 WⅠ 型微乳液体系中的界面膜[50,65]以及阳离子[64]和两性[66]表面活性剂在 WⅡ 体系中的界面膜。图 3.6 为 Eastoe 等得到的结果,它表明 $(2K + \bar{K})$ 为烷基碳数 $n\text{-}C$ 的函数。用(3.3.12)式和(3.3.15)式分别处理的一致性,也表明其正确性。所得到的值与目前的静态机械理论相一致[48],静态机械理论认为 K 在 $n\text{-}C^{2.5}$ 到 $n\text{-}C^3$ 间波动,而对 $\bar{K}$ 仅有很小的影响。

3.3.3 相行为

微乳液的溶解性和界面性质依赖于压力、温度以及组分的性质和浓度。稳态相图的绘制和多元体系(水(盐)-油-表面活性剂-醇)中各相的位置随各因素变化是非常重要的。一些相图能从含有的变化因素的个数上来预测。有几种不仅可以描述存在的单相区和多相区的界线,而且也可以表示出相间的平衡(连接线、连接三角、临界端点等)的表达方式。下面是对三相图和两相图的简单描

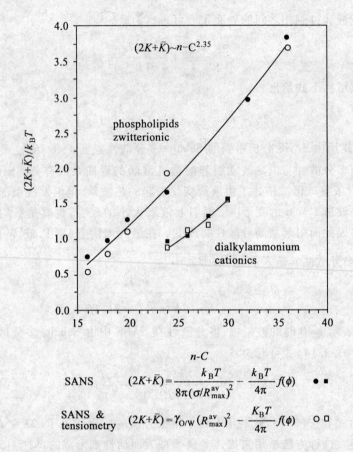

图 3.6 模硬度 $(2K+\bar{K})$ 与 Winsor Ⅱ 微乳的烷基链的总碳数的函数
(Eastoe et al. [64,66])

述以及组成的相律。

1. 相律

变量(或自由度)的数目与体系的组成和状态的关系由相律来确定。一般写作

$$F = C - P + 2 \qquad (3.3.16)$$

式中，F 是状态独立变量或自由度，C 是独立组分数，P 是体系中的相数。一个系统依照 F 等于 $0,1,2$ 等，分别叫做不变、单变和双变体系。例如，最简单的情况是一个有三个组分和二相的体系，当温度和压力恒定时，F 等于 1。这表示单

组分的物质的量或质量分数在一相中可变化,但在两相中的所有其他组分是固定的。一般来说,微乳液至少包含三个成分:油(O),水(W)和两亲分子(S),有时还有助表面活化剂(醇)及电解质调节体系的稳定性。这里,所有的体系都被认为是简单的 O-W-S 体系:即当用助表面活化剂时,油醇比保持恒定,且假设醇不与任何其他的组分相互作用。这样,混合体系可被当做一个三个组分体系。在压力恒定情况下,组分-温度相图见图 3.7。然而,绘制这样一张结构相当复杂的相图很耗费时间。为方便起见,通常通过研究特定的相分割来简化体系,即通过改变一个稳定体系的一个变量或者结合两个或更多的变量来减少变量的个数,然后再绘制三元相图和二元相图。

2. 三相图

在温度和压力恒定下,三组分微乳液的三个相图被简单划分为如图 3.8 所表示的两个或四个区域。在每种情况下,分界线以上的单相区的每个点对应着一个微乳液。在分界线以下的点对应含有微乳液与水的相或有机相或者两者相平衡的多相区,即 Winsor 型体系(参见 3.3.1)。

两相区内体系的全部组分(举例来说,在图 3.8(a)和图 3.8(c)中的点 o),将作为两相存在,表现为"连接线"的两端,也就是 m 和 n 两相。因此,在一个特定的连接线上的每点有着相同的共存相(m 和 n),但相对体积不同。两个共轭相有相同的组成($m = n$),对应于褶点 P(或临界点)。

如果三相共存(图 3.8(b)),即符合 WⅢ 体系在恒定的温度和压力下,依照相律三相体系所组成的三角形区域的组成不变,其边界是包围它的相邻二相的连接线。形成三角形内三个相的组成固定的区域,因此叫"连接三角"[26]。三角形内的任何点,例如图 3.8(b)的点 q 由划分为对应于三角形的顶点 A,B 和 C 的三相所组成。改变位置 q,点 A、B 和 C 组成不变,在组成三角形各处的组成中,A,B 和 C 的量在变但均由 A,B 和 C 组成。

3. 二元相图

如前所述,三元相图可进一步由固定一些参数(如:水 + 电解质→盐水,或水 + 油→水/油比)及结合两个变量而简化,也就是减少体系的自由度。这样,体系相图由相棱镜减少到平面区域便于研究。非离子型和阴离子型表面活性剂的拟二元相图见图 3.9 到 3.11。

图 3.9 是一个非离子型表面活性剂-水-油三元体系的概要相图。由于温度是非离子型表面活性剂的一个敏感因素,拟二元相图 $\phi_w = \phi_o$ 定义的平面部分,ϕ_w, ϕ_o 分别是水和油的体积分数。在恒定压力下,单相区域中定义一个体系需要两个独立变量($F = 2$),即温度和表面活化剂浓度。图 3.9(b) 中所表示的区域是三相平衡 W + M + O(M 为微乳液相)下,用来决定 T_L 和 T_U,分别地对应于

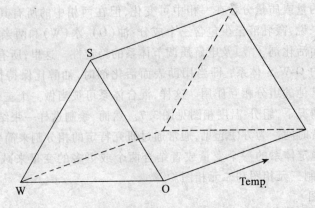

图 3.7 相棱镜,可描述恒定压力下三元体系的相行为

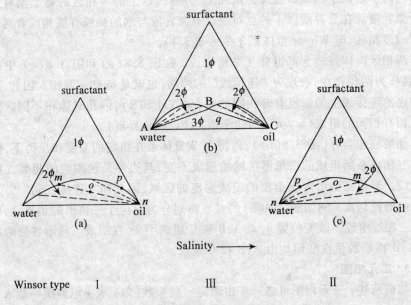

(a) Winsor Ⅰ 型;(b) Winsor Ⅱ 型;(c) Winsor Ⅲ 型

图 3.8 在不变的温度和压力下,由简单的水-油-表面活性剂体系形成的有着两相或三相区的三相图

低温和高温的界限。溶解等量的水和油所需表面活性剂的最小量,表示为 C_s^{*} [68]。C_s^{*} 越低,表示表面活性剂越有效。图 3.10 以非离子型表面活性剂-水-油三元体系为例说明第二种可能区域的测定,保持压力和表面活化剂浓度不变,而保留两个变量:温度和水/油比($\phi W-O$)。它展示了各种不同的表面活性剂

相与温度和水/油比的函数变化情况[68]。第三个例子(图 3.11)用到阴离子表面活性剂 Aerosol-OT,这个特别的拟二元相图将会在这本书中用到多处。在压力不变时,单相区域中定义三组分 W-O-S 体系时,固定表面活性剂浓度,所用两个变量是温度和水与表面活性剂摩尔比 $w(w = [水]/[表面活化剂])$。w 表示每个表面活性剂分子可溶解的水分子的数目,所以可用这个相图表示表面活性剂的效率,如微乳化效率。更多的关于 AOT 的相行为可参考文献[69,70]。

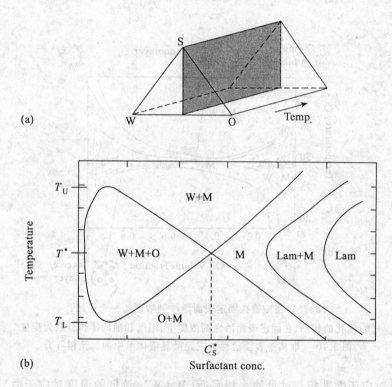

(a) 经过相棱镜在相等的水和油容量的区域;
(b) 简化的相图:温度和表面活性剂浓度 C_s 为变量
图 3.9 非离子表面活性剂形成三相微乳液体系的拟二元相行为

T_L 和 T_U 分别是在相平衡 W + M + O 时的低温和高温界限。T^* 代表三相区的三角形等边时的温度,也就是说,中相微乳液在该温度含有等量的水和油。这种情况也被称为平衡点。C_s^* 是在平衡的情况下,中相微乳液中表面活性剂浓度。"Lam" 表示层状液晶相。参见 Olsson 和 Wennerström[68]。

请注意微乳液相(M)在不同区域中有各种不同的微细结构,在液相中,较高

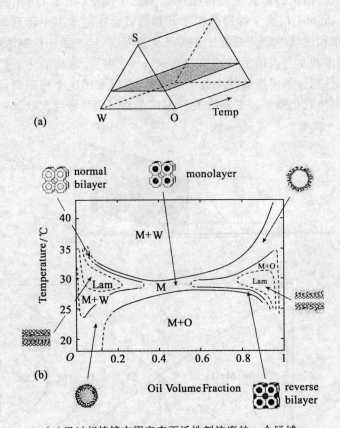

(a) 经过相棱镜在固定表面活性剂浓度的一个区域；
(b) 简化的相图，在固定表面活性剂浓度下，温度和油的体积分数为变量
图 3.10 非离子表面活性剂三元微乳液体系中的二元相行为

的温度对应于微乳相与过量的水相平衡(M+W)，较低的温度对应于微乳相与过量的油相平衡(M+O)。在中间的温度都会有一个层状相分别与较高的水和较高的油稳定共存。参见 Olsson 和 Wennerström[68]的论文。

在固定表面活性剂浓度和压力下，以水/表面活性剂摩尔比 w 及温度为变量。图中数字为 n，有圈的数字对应低温（溶解）边界，T_L。没有圈的数字对应高温（薄雾）边界，T_U。单相微乳液区域位于 T_L 和 T_U 之间。在 T_L 以下，体系为微乳液相和多余的水相平衡（WⅡ型）。在 T_U 以上，体系由微乳液相分成富含表面活性剂的相和多余的油相平衡。参见 Fletcher et al[16]的论文。

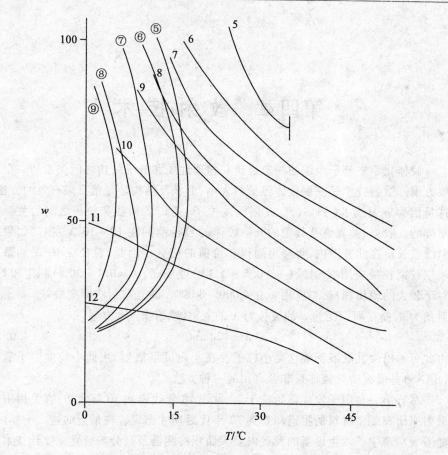

图 3.11 阴离子表面活性剂 Aerosol-OT(AOT) 在各种不同的直链烷烃溶剂中形成三元微乳液体系中的拟二元相图

参考文献(略)

第四章 散射技术

胶体化学中分子组织的测定是研究物理性质和分子结构之间关系的一个重要方面。散射技术是获取胶体粒子的大小、形状及结构定量信息最有效的方法，这是因为散射技术(例如，光、X 射线或中子)是以入射线和粒子之间的关系为基础的。胶束、微乳液及其他的分散胶体的粒径范围在 $10 \sim 10^4$ Å 左右，如果入射线波长落在此范围内，将会得到很有价值的信息。因此，粒径在 10^2 Å 的微乳液或胶束粒径应用 X-射线($\lambda = 0.5 \sim 2.3$ Å)或中子($\lambda = 0.1 \sim 30$ Å)测定较好。对于较大的胶体颗粒，则应用 λ 在 $4\,000 \sim 8\,000$ Å 的光散射法测定最好。对于衍射角为 θ，晶格面间距为 d 的波长为 λ 的入射波，有 Bragg 方程：

$$\lambda = 2d\sin\theta \qquad (4.0.1)$$

由此可看出微乳液液滴等这类小粒子会在小角度下散射，用此小角度中子散射(SANS)是研究微乳液体系非常有用的一种方法[1]。

尽管第一台中子发生器始建于 20 世纪 40 年代末或 50 年代初，有关用中子散射来研究凝聚材料的报道却直到 70 年代后期才出现。在最近的这二十年内，随着大功率中子发生设备的发展以及大面积检测器及高分辨分光光度计技术方面的进步，SANS 已经变为一种更加实用的技术，特别是成功地用于胶束、微乳液和液晶的构成方面的研究。SANS 是较新的也是测试分子聚集结构的一个强有力的工具。

在下述几个部分中我们将就中子散射理论和 SANS 数据分析方法做一个综述。

4.1 背景知识简介

4.1.1 中子

中子是不带电的(电中性)，质量约为 1.675×10^{-27} kg (是电子的 1 839 倍)，磁矩为 -1.913 核磁子，自旋量子数为 1/2 的亚原子粒子。当中子被限制在原子核内部时是稳定的，而当其以自由粒子形式存在时则平均寿命仅在 1 000

s 左右。中子和质子几乎占了整个原子核的质量,因此它们都被称为核子。依照波长和能量对中子进行分类,短波长($\lambda \sim 0.1 Å$)的中子称为超热的,较长波长($\lambda \sim 10 Å$)的中子称为"热的"或"冷的"。通过在对中子发生设备如反应堆和散裂源的调节可以控制中子的波长至预期的范围内。

中子与物质通过强、弱电磁和重力的交互作用互相影响,然而,正是由于这两个力的相互作用(强的短程核力和它们的磁矩)使中子散射成为探索凝聚态物质结构的独特技术。在微观水平上对结构和动力学研究方面,中子较其他散射技术的显著优越性在于:

- 中子是不荷电的,因而它们可以穿透大块的物质。它们靠强核力与研究材料的核子相互作用。
- 中子的磁矩与原子范围内磁性空间变化相对应。因此它们很适合于研究磁性结构、自旋体系的波动和激发(中子与原子核发生弹性碰撞后使其运动方向发生变化,即使其磁矩发生变化,译者注)。
- 中子的能量和波长通常可以同时满足凝聚态物质结构及激发对能量和波长范围的要求。从下面德布罗意方程可知,波长 λ 与中子运动速度有关:

$$\lambda = \frac{h}{mv} \qquad (4.1.1)$$

式中,h 是普朗克常数($6.63 \times (10^{-34} J s)$,$v$ 是粒子的速度。

相关的动能为

$$E = 1/2 mv^2 \text{ 或 } E = \frac{h^2}{2(m\lambda)^2} \qquad (4.1.2)$$

由于波的能量和波长取决于中子的速度,因而有可能通过飞行时间技术来选择特定的中子波长。

- 中子的存在不会对检测体系产生明显的干扰,因此中子散射实验的结果可以被清楚地解释。
- 中子无破坏性,甚至可用来检测精细的生物材料。
- 中子的高穿透力使其可以检测大体积的材料,并且不易受样品环境的影响(例如高压、温度和磁场)。
- 中子散射来自于材料原子核的互相影响而非电子云。这也就意味着原子的散射力(横截面)与原子序数关系不大(原子序数是原子中质子的数目或电子的数目,因为原子是呈中性的),不像 X 光和电子其散射能力随原子数成比例地增加。因此,通过中子散射,可将较轻的原子如氢原子同较重的原子区别开。与此类似,在周期表中相邻元素的中子散射通常有实质性的不同因而可以被区分。中子散射对核的依赖性甚至使得同一原子的各同位素也可通过散射区分开来,这样就可以通过使用同位素对材料的各部分作标记来对其进行考察。

4.1.2 中子源

中子束的生产有两种方式:基于反应堆中子源的核裂变,或基于加速器中子源的散裂。下面就这些工艺做一个简介,其中特别参考了目前世界上最著名的两个中子来源地——法国 Grenoble 的 Institut Laue Langevin(ILL)[2],和英国 Didcot 的 Rutherford Appleton 实验室的 ISIS 设备[3]。

- 基于反应堆的中子源:传统的中子是通过可产生高亮度中子的优化的核反应堆裂变产生的。在该过程中,铀 235 核在吸收了热的中子后裂变为碎片的同时蒸发出一个高能(MeV)的稳定中子流(因此有术语"稳态的"或"连续的"中子源)。在周围的减速器中高能量中子热能达到兆电子伏后,发射出一个宽范围波长的束流。使中子流与"热源"达到热平衡(在 ILL 中热源通常为自加热的 2 400 K 的石墨堆)即可使中子能量分布升高,使用"冷源"如 25 K 液态氘即可使中子流能量分布降低[4]。最终的中子的 Maxwell 能量分布具有减速器的特征温度(图 4.1(a))。波长选择一般是通过一个晶体单色器的 Bragg 散射,也可通过机械斩波装置调节其速度从而控制其波长。这样就制得了散射实验用的狭窄波长分布的高质量、高流量中子束流。今天世界上最大功率的反应器中子源是 ILL 的 58 MW HFR(高流量反应堆)。

- 基于加速器的脉冲中子源:用由高能加速器获得的高能粒子(如 H^+)轰击重金属靶(如 U,Ta,W)从而获得中子流(一般称为散裂)。粒子加速的方法可定时输出强的高能质子流,进而产生中子脉冲。每个有用中子的散裂较裂变释放的热较少(与典型的 190 MeV 裂变相比较,每个散裂的中子为 30 MeV)。低热耗散意味着脉冲源可输送高亮度中子(亮度超过大多数先进的稳态中子源),同时在金属靶上产生较少的热量。目前世界上功率最大的散裂中子源是 ISIS 设备。它基于一个 200 μA,800 MeV,50 Hz 下操作的质子同步回旋加速器和一个钽(Ta)靶,该金属靶在每个入射质子的轰击下可产生 12 个中子。

在 ISIS 中,产生能量足以引起充分散裂的粒子主要包括以下三个阶段(图 4.2):

(1) 从氢气中产生 H^- 离子(带两个电子的质子)并在预注入柱中对其加速使其能量达到 665 keV。

(2) 在含有四个加速管的线形加速器中将 H^- 离子加速至 70 MeV。

(3) 在同步回旋加速器中加速,同步加速器直径 52 m,每一脉冲可将 2.8×10^{13} 个质子加速到 800 MeV。在进入同步加速器时 H^- 先经过一个极薄的(0.3

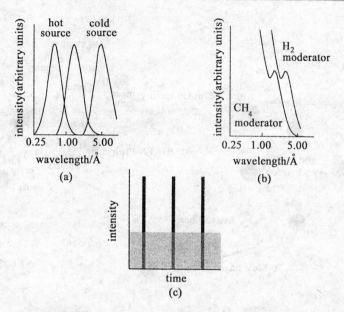

图4.1 （a）是来自反应堆一个典型的中子波长分布,如图所示分别为来自于热源（2 400 K）,温热源和冷源（25 K）的三个峰。峰分布经归一化处理使 Maxwell 分布为1。

（b）来自一个脉冲散裂源的典型波长分布图。减速剂 H_2 和 CH_4 温度分别为 20 K 和 100 K。峰形具有 Maxwell 分布并含有一高能量"减慢"成分和一个热能化成分。峰分布经归一化处理。

（c）稳态源（灰色部分）和脉冲源（黑色部分）的中子流随时间变化图。稳定源中子流例如 ILL 系时均流量较高,而脉冲源中子流如 ISIS 则具有较高亮度（未按坐标尺画出）[3]。

μm）氧化铝金属薄片,将 H^- 的电子除去而转变成质子。在同步加速器加速中（约需 10 000 r/min）,每一转都是由电磁场加速的,800 MeV 的质子束被从同步加速器中引出后轰击金属靶而产生中子。

质子束与靶原子碰撞产生大量的中子,与裂变相同,它们也必须通过减速材料才能得到预期的波长范围。靶周围的氢可作减速剂。因为氢具有大的非弹性散射截面,在与氢核反复碰撞后中子的运动速度被减缓。减速剂的温度决定了产生的中子束的峰分布,可以通过不同的实验手段进行控制（图4.1（b））。ISIS 所使用的减速剂为室温的水（316 K, H_2O）,液态甲烷（100 K, CH_4）和液态氢（20 K, H_2）。

由脉冲源产生的中子束的特性与反应堆产生的中子有很大不同（图4.1

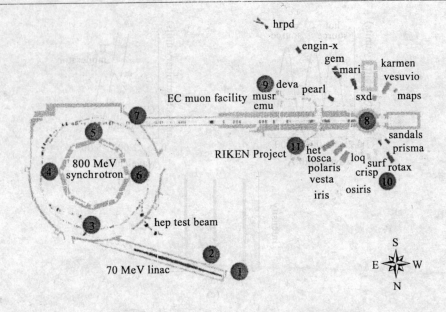

1. Ion source and per-injector
2. 70 MeV linear accelerator
3. 800MeV synchrotron injection area
4. Fast kicker proton beam extraction
5. Synchrotron south side
6. Synchrotron west side
7. Extracted proton beam tunnel
8. ISIS target station
9. Experimental hall, south side
10. Experimental hall, north side
11. RIKEN superconducting pion decay line

图 4.2　Rutherford Appleton 实验室散裂脉冲中子源示意图（英国，Didcot，ISIS）光束自 ISIS 靶发射出来，输送出"白色"脉冲中子（中子具有宽的波长分布）到设备 18 [3]

相对反应堆的中子束仍然较低。然而合理地使用飞行时间技术（TOF）通过对中子亮度的测定可以弥补这个缺陷，通过使用飞行时间技术可以直接分析出"白色"中子束（具有宽的波长分布，与白光和单色光之间的关系类似，译者注）中每个中子的波长及能量。

4.1.3　SANS 设备

在中子散射实验中，设备可以通过波向量 Q 计量散射的中子数，而 Q 与散射角 θ 和波长 λ 有关（见 4.4 节）。对于弹性散射（散射中子与入射中子具有相同的能量），这就相当于利用衍射仪对磁矩变化进行测定（原子核的运动方向改变，会引起其磁矩变化，而发生弹性散射时这种运动变化完全是由中子碰撞引起的，译者注），这样就可以获得体系中关于原子核的尺寸复杂程度等空间分布的信息,这些体系可以是小的单元晶格或玻璃、液体等无序体系直至"大规模"的表面活性剂聚集体及聚合物。另一方面,分光光度计可用来对中子与样本相互作用过程中的能量损失(或增益)进行分析(如非弹性散射会引起能量损失),用这些数据可以推知样本的动态行为。

在 D22 设备中,通常使用一个单波长的中子束,可以通过斩波装置调节中子流速获得单波束。相比之下,LOQ 使用"白色"的即多波长的中子束。散射光束的能量分析是测定总飞行时间(TOF)而获得的,如可以测定从中子源到样本飞行所需要的时间。由于具有不同的波长分布,LOQ 和 D22 的探测仪器也有很大的不同。当入射波长给定时,必须测定不同角度的散射强度以获得预期的 Q 范围。D22 使用可移动的探测器,改变样本到探测器的距离即可达到上述要求。而对于 LOQ,探测器的位置是固定的,中子脉冲波的变化可以通过 TOF 测定。图 4.3 和图 4.4 给出了两种装置的示意图,关于其技术细节可以参考文献[2,3,5]。

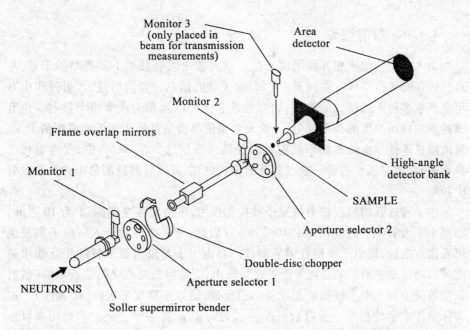

图 4.3 LOQ 装置示意图(ISIS, Didcot, U. K [2])(在与样本发生作用后(典型的中子向样本的流速 $=2\times10^5 cm^{-2}\cdot s^{-1}$),中子束通过一个真空管,其中有一个氙气填充的探测器(有效面积 $64\times64 cm^2$,像素大小 $6\ mm\times 6\ mm$)探测器与样本的距离为 4.5 m。入射波长范围约为 2.2~10Å,散射角 <7°,有效 Q 范围为 0.009~0.249Å^{-1})

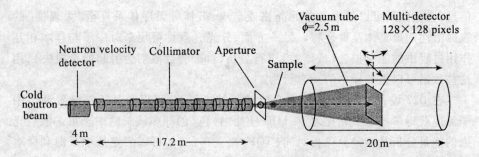

图 4.4 D22 装置示意图(ILL, Grenoble, France)[1](通过样本的最大中子流速为 $1.2 \times 10^8 cm^{-2} \cdot s^{-1}$。在所有的小角度散射装置中,D22 拥有最大面积的多极检测器(^{3}He)(有效面积为 96 cm×96 cm, 像素大小为 7.5 mm×7.5 mm)。该检测器被套入一个 2.5 m 宽、20 m 长的真空管,样本到探测器的距离可在 1.35~18 m 之间调节;检测器可在径向移动约 50 cm,并可绕轴旋转以减小视察。因此 D22 检测范围为对于 $\lambda = 2.6 Å$,Q 范围可达到 $1.5 Å^{-1}$(对于 $\lambda = 4.6 Å$,$0.85 Å^{-1}$,$\Delta\lambda/\lambda = 5\% \sim 10\%$))

4.1.4 散射理论

由入射波与物质相互作用而产生的散射波之间会发生不同类型的干涉,从这些干涉图可推知样本空间及(或)时间关联的信息。如前所述,入射过程中有可能产生多种不同模式的散射,如弹性或非弹性以及相干或非相干散射。由有序排列的核所产生的相干散射,可能发生提供结构信息的相增干涉或相消干涉。而由随机事件产生的非相干散射可以提供动态信息。在 SANS 中,仅考虑相干弹性散射,而非相干散射一般以背景的形式出现,可以容易地测量出并从总散射中扣除。

中子通过强的短程核力与原子核发生作用,该力的存在范围约为 10^{-15} m,即远小于入射中子的波长($\sim 10^{-10}$ m)。因此,每一个核对于入射中子束来说可看做点散射,该中子束可看做平面波。自由中子与束缚原子核的相互作用强度可以通过原子的散射长度 b 而被定量求出,而散射长度也受同位素影响(散射长度数值上等于中子波矢量相位移的相反数,是原子密度的表征,译者注)。在实际操作中平均相干中子散射密度 ρ_{coh},简写做 ρ,是一个合适的参数用来对体系中不同组分的散射效率进行定量。ρ 表示单位体积物质的散射长度,是整个分子体积 V_m 内所有原子贡献的加合:

$$\rho_{coh} = \frac{1}{V_m}\sum_i b_{i,coh} = \frac{DN_a}{M_w}\sum_i b_{i,coh} \quad (4.1.3)$$

式中,$b_{i,coh}$ 是质量密度为 D,分子量为 M_w 的分子中第 i 个原子的相干散射长度,

N_a 为阿佛加德罗常数。一些有用的原子散射长度值列于表 4.1，一些分子的散射密度也列于该表中[6]。氢与氘的 b 值有明显不同，这可用于对比变异技术来检测不同区域分子组装情况，比如我们可以"看见"含有质子的烃类化合物在重水中的溶解情况。

表 4.1　　　　　　　一些原子的相干散射长度值 b[6]

原子核	$b/(10^{-12}\text{ cm})$
^{1}H	-0.374 1
^{2}H (D)	0.667 1
^{12}C	0.664 6
^{16}O	0.580 3
^{19}F	0.565 0
^{23}Na	0.358 0
^{31}P	0.513 1
^{32}S	0.284 7
Cl	0.957 7

a 值由氘化表面活性剂离子计算得到；无 Na 阳离子，烷基链为氘化的。

在中子散射实验中，在不同的散射角 θ 下测出散射波的强度 I，并且在 SANS 实验中 θ 一般远小于 10°。图 4.5 给出了 SANS 实验的示意图，在该图中入射中子束被样本散射，入射波为一平面波，其振幅可写作[7]

$$A_{\text{in}} = A_o \cos(\underline{k}_o \cdot \underline{R} - \Omega_o t) \tag{4.1.4}$$

式中，A_o 为初始振幅，$\underline{k}_o$ 为大小为 $\frac{2\pi}{\lambda}$ 的波向量，$\underline{R}$ 为位置向量，Ω_o 是频率，t 为时间。在静态实验中，分子间相对运动被忽略，没有时间相关性。如果再考虑复合振幅，则（4.1.4）式可简化为

$$A_{\text{in}} = A_o \exp(i\underline{k}_o \cdot \underline{R}) \tag{4.1.5}$$

当这个中子波碰撞一个原子时，部分中子沿散射中心呈球形散射。

$$A_{\text{sc}} = \frac{A_o b}{r} \exp(i\underline{k}_o \cdot \underline{R}) \tag{4.1.6}$$

式中，b 为散射长度，r 为两散射点间的距离（图 4.6(a)）。如果原子不在原点而在位置向量 $\underline{R}$ 处，$\underline{k}_s$ 方向上的散射波将会关于入射波发生 $\underline{Q} \cdot \underline{R}$ 的相转移（图 4.6(b)）。$\underline{Q}$ 为散射向量，由下式与散射角 θ 关联：

$$\underline{Q} = \underline{k}_s - \underline{k}_o \tag{4.1.7}$$

向量 $\underline{Q}$ 的大小可由余弦法则算出

$$Q^2 = k_o^2 + k_s^2 - 2k_o k_s \cos\theta \tag{4.1.8}$$

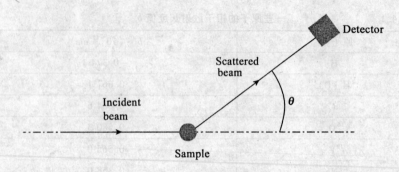

图 4.5　小角散射实验装置示意图(样本到检测器的距离为 1~20 m；散射角 $\theta < 10°$)

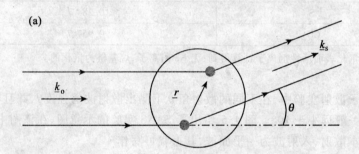

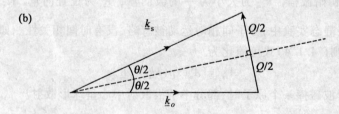

(a) 两散射点间的相差异，两散射点间空间位置向量为 r。入射波及散射波的波向量分别为 k_o 和 k_s。对于弹性散射 $|k_o| = |k_s| = 2\pi n/\lambda$。(b) 散射向量的计算 $Q = k_s - k_o$，大小为 $Q = (4\pi/\lambda)\sin(\theta/2)$

图 4.6　散射实验中的几何关系

对于相关弹性散射，$|\underline{k}_o| = |\underline{k}_s| = \dfrac{2\pi n}{\lambda}$，式中，$n$ 为媒介的折光指数，对于中子该值约为1，因此 $|\underline{Q}|$ 可由简单的几何计算得到

$$|\underline{Q}| = Q = 2|\underline{k}_o|\sin\dfrac{\theta}{2} = \dfrac{4\pi}{\lambda}\sin\dfrac{\theta}{2} \qquad (4.1.9)$$

Q 的数量单位为长度的倒数，一般为 Å^{-1}。该值与样本的空间特性有关，一般大结构在较低的 Q 值（及角度）、小结构在较高的 Q 值下发生散射。

相应地，对于距离原点位置向量为 $\underline{R}$，角度为 θ 的散射波的振幅为

$$A_{sc} = \dfrac{A_o b}{r}\exp[i(\underline{k}_o r - \underline{Q}\cdot\underline{R})] \qquad (4.1.10)$$

(4.11)式仅适用于考虑两个散射点的情况，而实际中对于更一般的多原子分子，总的散射振幅可写作下式：

$$A_{sc} = \dfrac{A_o}{r}\exp(i\underline{k}_o r)\sum_i b_i \exp(-i\underline{Q}\cdot\underline{R}_i)] \qquad (4.1.11)$$

对于一定区间内的 SANS 及 Q 范围（距离约为 10~1 000Å，散射向量 Q 约为 0.006~0.6Å^{-1}），样本可以看做是在一连续介质中不连续分散的粒子，散射由散射密度 ρ 控制：

$$\rho(\underline{R}) = \dfrac{1}{v}\sum_j b_j\delta(\underline{R} - \underline{R}_j) \qquad (4.1.12)$$

式中，加和是对整个体积 V 而言，该体积与分子内原子间距相比较大但与实验的分辨率相比则较小。因此散射振幅是受照射体积 V 中密度的傅里叶变换：

$$A_{cs}(\underline{Q}) = \int_v \rho(\underline{R})\exp(-i\underline{Q}\cdot\underline{R})d\underline{R} \qquad (4.1.13)$$

因为对相位移不敏感，所以发射检测器不能测量振幅值，但可以检测散射强度 I_{sc}（或功率通量），该值是振幅模的平方：

$$I_{sc}(\underline{Q}) = \left(\left|A(\underline{Q})\right|^2\right) = \left(A(\underline{Q})\cdot A^*(\underline{Q})\right) \qquad (4.1.14)$$

对于一个包含 n_p 个相同粒子的体系，(4.15)式变成[8]

$$I_{sc}(Q) = n_p\left(\left(\left|A_{sc}(Q)^2\right|\right)_o\right)_s \qquad (4.16)$$

在这里体系对所有方向 o 及形状 s 取平均值。

因此，在两个参变量 θ 和 λ，与距离的倒数 Q 之间存在着一个简单的关系式（(4.19)式），这里 Q（通过(4.13)式）与样本中关联两个原子核散射点位置的 r 有关。这些参数都与散射强度 $I(Q)$ 相关（(4.1.15)式），而该值可由 SANS 实验直接测定，这样就可能得到样本的粒子间及粒子内部的结构信息。

4.2 胶束聚集体的中子散射

对于单分散的半径为 R,体积为 V_p,数密度为 $n_p(cm^{-3})$ 和相干散射密度为 ρ_p 的分散于密度为 ρ_m 的介质中的均匀球形粒子,归一化的 SANS 强度 $I(Q)$ (cm^{-1}) 可写作[9]

$$I(Q) = n_p \Delta\rho^2 V_p^2 P(Q,R) S(Q) \tag{4.2.1}$$

式中,$\Delta\rho = \rho_p - \rho_m (cm^{-2})$,在 (4.2.1) 式中的前三项与 Q 无关,表示散射强度的绝对值。对于比例因子 S_F 其定义如下:

$$S_F = n_p(\rho_p - \rho_m)^2 V_p^2 = \phi_p \cdot \Delta\rho^2 \cdot V_p \tag{4.2.2}$$

式中,ϕ_p 是粒子的体积分数。比例因子是对描述 SANS 数据的模型的有效性及一致性的量度。例如由模型拟合得到的 S_F 值可与由 (4.2.2) 式算得的计算值相比较。(4.2.1) 式中的后两项是关于 Q 的函数。$P(Q,R)$ 是由粒子内散射获得的单粒子形状因子。它用来描述由于离子形状及尺寸变化造成的散射角分布。$S(Q)$ 是由粒子间相互作用得到的结构因子。为了更好地理解每一项的影响,在图 4.7 中给出两个分别为均匀球形粒子互斥及相互吸引情形的散射图[8]。该图说明了 $P(Q)$ 和 $S(Q)$ 的相互关系对最终总散射强度 $I(Q)$ 的影响。对这些散射函数简要讨论如下。

4.2.1 单粒子形状因子 $P(Q)$

通过 $P(Q)$ 函数可以获得粒子尺寸及形状的信息,球形粒子的形状因子 $P(Q,R)$ 的近似示意图见图 4.7。一般来说,该曲线呈下降趋势,尽管在高 Q 时,经常会出现极大和极小值。在 $Q=0$ 时,函数 $P(Q)$ 一般定义为 1.0。对于很多不同形状的粒子,都已经获得了 $P(Q)$ 函数的一般表达式,如均匀球形、球壳形、圆柱形、同心圆柱形及圆盘形[7]。对于半径为 R 的球形有

$$P(Q,R) = \left[\frac{3(\sin QR - QR\cos QR)}{(QR)^3}\right]^2 \tag{4.2.3}$$

对于特定的体系如微乳液,需要引入多分散函数来解释粒径分布,对于球形液滴,这种粒径分布可用 Schultz 分布函数 $X(R_i)$ 表达[10,11]。定义 R^{av} 为平均半径,$\sigma = \dfrac{R^{av}}{(Z+1)^{1/2}}$ 为均方根偏差,Z 为宽度参数。$P(Q,R)$ 的表达式如下:

$$P(Q,R) = \left[\sum_i P(Q,R_i) X(R_i)\right] \tag{4.2.4}$$

4.2.2 结构因子 $S(Q)$

粒子间结构因子 $S(Q)$ 与体系内相互作用的类型有关,如吸引、排斥及空间

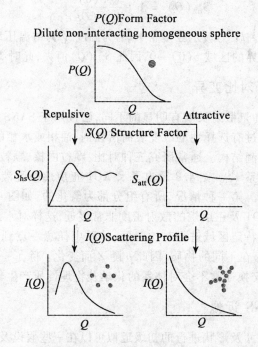

图 4.7　相互吸引及相互排斥的均匀球形粒子形状因子 $P(Q,R)$ 及结构因子 $S(Q)$ 及其对散射强度 $I(Q)$ 贡献的示意图[8]

位阻效应。对于球形粒子,如果其相互作用较弱则可初级近似为硬球势能模型 $S_{hs}(Q)$,如式(4.2.5)所示[12]:

$$S_{hs}(Q) = \frac{1}{1 - n_p \cdot f(R_{hs}\phi_{hs})} \quad (4.2.5)$$

式中,$R_{hs} = R_{core}^{av} + t$ 为硬球半径(t 为烃层厚度),$\phi_{hs} = \frac{4}{3}\pi R_{hs}^3 n_p$ 为硬球体积分数,散射强度表达式(4.2.1)式即可写成

$$I(Q) = \phi_p \Lambda \rho^2 V_p \left[\sum_i P(Q,R_i) X(R_i)\right] S(Q,R_{hs},\phi_{hs}) \quad (4.2.6)$$

如图 4.7 所示,在较低的 Q 值时,$S_{hs}(Q)$ 可以减小散射强度并在 $I(Q)$ 分布图中在 $Q_{max} = 2\pi/D$ 处有一个峰,D 为样本中原子的平均最小距离。对于稀的无相互作用的体系,$\phi_{hs} \to 0$,因此结构因子消失,如 $S(Q) \to 1$。对于相互作用体系,减小 $S(Q)$ 的一个有效方法是将体系稀释[13],对于带电粒子可以加电解质[14]。

对于需要考虑相互吸引作用的体系,尤其是二元相图中浊点及相分离区附近,可以使用 Ornstein-Zernike(OZ)表达式作为结构因子函数[8]:

$$S_{OZ}(Q) = 1 + \frac{S(0)}{1 + (Q\xi)^2} \qquad (4.2.7)$$

式中，$S(0) = n_p k_B T \chi$，k_B 为玻尔兹曼常数，T 为温度，χ 为恒温压缩系数，ξ 为关联长度。在远离相边界的区域 $S(0) \to 0$，因此 $S_{OZ}(Q)$ 消失，此时 $S(Q) \to 1$。

4.2.3 中子对比变异

如前所述，氢和氘散射长度有明显的不同，已被用于 SANS 实验来考察界面的结构与组成。通过有选择性的改变表面活性剂、油相或水相的散射长度密度，可以考察微乳液液滴结构。通常研究三种对比：颗粒内核、颗粒外壳及小液滴的单独或同时拟合体系[15]。图 4.8 示出了水-AOT-正庚烷微乳液体系的散射长度密度图的三种对比。第一种情况，所有组分都为氢化的，如图 4.8(a) 所示。如表 4.2 所示，水、AOT 及正庚烷的散射密度非常接近，这样对水或(和)烃链选择氘化就可对体系内特定区域进行对比匹配，体系中任意一点到液滴中心的距离为 Z，由于体系内存在不同的物质，因此 ρ 随 Z 而变化。除了一些精细的效应如氢键等，该同位素交换通常不会对体系的化学及物理性质产生较大的影响。

4.2.4 SANS 近似

对于粒子的尺寸及形状进行的初级近似可以在一些假设及近似的基础上由 $I(Q)$ 与粒子半径(或厚度)之间简单的关系式获得。

Guinier 近似

SANS 分布图 $I(Q)$ 对于不同的粒子形状非常敏感。特殊地，Guinier 近似将散射图中低 Q 区域与粒子的回转半径 R_g 联系起来。对于稀溶液体系，在低 Q 区间(Guinier 区间)，单个粒子形状因子 $P(Q,R)$ 可简化成[16]

表 4.2　　　　　　一些分子在 25°C 下的相关散射密度[6]。

	分子	$\rho / (10^{10} \text{ cm}^{-2})$
水	H_2O	-0.560
	D_2O	6.356
庚烷	C_7H_{16}	-0.548
	C_7D_{16}	6.301
AOT	$(C_8H_{17}COO)CH_2CHSO_3^-(Na^+)$	0.542
	$(C_8D_{17}COO)CH_2CHSO_3^-(Na^+)$	5.180^a

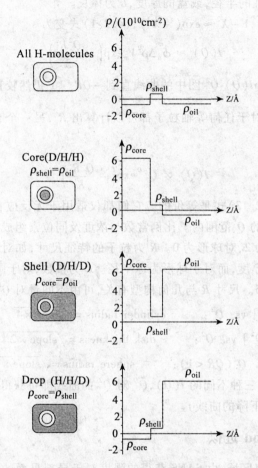

图 4.8 通过对比变化对水-AOT-正庚烷微乳液体系结构的分析散射长度密度 ρ 依赖于到液滴中心的距离 Z

$$P(Q,R) = 1 - \frac{Q^2 R_g^2}{3} \tag{4.2.8}$$

式中,R_g 为对整个粒子体积的平均半径取均方根,与粒子形状有关:

- 球形或圆柱形 $\qquad R_g = \left(\frac{3}{5}\right)^{1/2} R \tag{4.2.9}$

- 薄盘形 $\qquad R_g = \dfrac{R}{4^{1/2}} \tag{4.2.10}$

- 长棒形 $\qquad R_g = \dfrac{L}{12^{1/2}} \tag{4.2.11}$

式中,R 为球或圆柱的半径,或盘的厚度,L 为棒长。

假定 $S(Q)=1$,$1-X^2 \approx \exp(-X^2)$,(4.2.1)式变为

$$I(Q) \approx \phi_p \Delta\rho^2 V_p \exp\left(-\frac{Q^2 R_g^2}{3}\right) \tag{4.2.12}$$

Guinier 图如 $\ln I(Q)\text{-}Q^2$ 图中的曲线直到 $-QR_g < 1$ 仍然较直,该直线部分的斜率为 $-\frac{R_g^2}{3}$,因此对于任何等轴粒子都可以计算出 R_g,另一个有用的 Guinier 近似表达式为[7,17]

$$I(Q) \propto Q^{-D} \exp\left(-\frac{Q^2 R^2}{K}\right) \tag{4.2.13}$$

(4.2.12)式与(4.2.13)式是等价的。它们都仅适用于无反应粒子(如 $S(Q) \to 1$),且必须在有限的 Q 范围内。比例常数与浓度及同位素组成有关。指数 D 对于圆柱为 1,盘状为 2,对球形为 0。R 为粒子的特征尺寸,如对于圆柱为横截面半径,对于盘状为厚度,而对于球形为球半径。K 为整数,对于圆柱形、盘形及球形分别为 4,12 和 5。尺寸 R 与几何构型有关,可由不同量对 Q^2 作图而获得。

- $\ln[I(Q) \cdot Q]$ vs. Q^2:　　cylinder radius $= \sqrt{\text{slope} \times 4}$ (4.2.14)
- $\ln[I(Q) \cdot Q^2]$ vs. Q^2:　　disk thickness $= \sqrt{\text{slope} \times 2}$ (4.2.15)
- $\ln[I(Q)]$ vs. $Q^2 (QR<1)$:　　sphere radius $= \sqrt{\text{slope} \times 5}$ (4.2.16)

因此通过对比三种不同的 $I(Q)$,Q^D 对 Q^2 的曲线即可获得最可能的粒子形状(例如一条线性下降的曲线)。

4.2.5 Porod 近似

在较高的 Q 值区域,由 SANS 获得的强度对于局部界面的散射比对整个粒子间结构的散射还要敏感,因此 $I(Q)$ 与整个界面面积 S 都有关,其渐进线强度(见图 4.9)可由 Porod 近似分析得到[18,19]

$$I(Q) = 2\pi\Delta\rho^2 \left(\frac{S}{V}\right) Q^{-4} \tag{4.2.17}$$

式中,S/V 为单位体积溶液内的总的界面面积(cm^{-1})。Porod 方程仅适用于平滑界面且 Q 范围 $\gg 1/R$(Porod 区间)。假定所有的表面活性剂分子处于界面,则平均表面活性剂亲水基面积 a_s 可用下式估算出

$$a_s = \left(\frac{S/V}{N_s}\right) \tag{4.2.18}$$

式中,N_s 为表面活性剂分子的数密度(如表面活性剂浓度×阿佛加德罗常数)。Porod 近似也可以被用来估算粒径[8]。对于单分散的半径为 R 的球形而言,

$[I(Q) \cdot Q^4]$ 对 Q 作图在 $Q \approx 2.7/R$ 处出现第一个极大值,在 $Q \approx 4.5/R$ 处出现极小值(见图 4.9)。

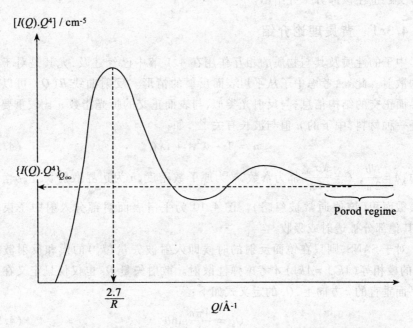

图 4.9 对近单分散的球形体系的 Porod 曲线示意图(详细介绍见正文)

Guinier 和 Porod 给出了简单的近似关系式,用来对胶体粒子的尺寸及形状进行初步估计。然而它们都仅适用于稀的非反应体系。如前一节所述,对溶液稀释或外加电解质都可对亲水基间的相互作用产生屏蔽,这样在低 Q 范围内 $S(Q)=1$ 的假设可以成立,因此可以运用 Guinier 近似。对于微乳液体系这些条件不一定成立,因为在稀释或外加电解质的情况下微乳液结构变化会使其稳定性遭到破坏。在这种情况下,要获得聚集体的尺寸及形状信息,需要使用更复杂的数学模型对 SANS 实验数据进行拟合,如由多分散球形液滴获得的并在本章中介绍的模型。关于这些模型的详细介绍见文献[9,20]。

4.3 中子反射

中子反射(NR)是一种有效且可靠的方法,因为它可以被直接用来测定表面过剩,并且可以用来对界面特性进行描述。与 SANS 一样,NR 的缺点也是操作费用昂贵,并且有时还需要氘化的溶剂和/或表面活性剂。表面张力法也是一种测表面过剩的可行方法,但该法只能通过 Gibbs 方程计算间接得到表面过剩量(见 2.1.2)。与 NR 相关的材料及定义已经在小角度散射中叙述过,必要的参考文献也在该部分一并给出。

4.3.1 背景理论介绍

中子的性质及其与物质的相互作用在 4.1 节中已经述及,尤其是对于小颗粒的散射。此处,考虑中子从平坦表面反射的情形。反射曲线 $R(Q)$ 可以提供与界面正交的结构信息,与反射光类似,与表面正交的折光指数 n 比较重要。对于任一种材料,中子的 n 值与波长有关[21],如:

$$n = 1 - \lambda^2 A + i\lambda C \tag{4.3.1}$$

式中,$A = \dfrac{Nb}{2\pi}$,$C = \dfrac{N\sigma_{abs}}{4\pi}$ 都为常数,N 为原子数密度,b 为边界散射长度,σ_{abs} 为吸附横截面积(该值通常被忽略)。图 4.10 为平滑表面对部分入射中子束的反射,其他部分被透射或吸收。

对于 SANS,则只在镜面反射的时候即入射波矢量(k_0)的模和反射波矢量(k)的模相等($|k_0| = |k|$)才考虑弹性散射。散射矢量 Q_z 也仅仅只定义在与入射平面垂直的 z 方向上,Q_z 的定义式如下:

$$Q_z = \dfrac{4\pi n}{\lambda}\sin\theta \tag{4.3.2}$$

式中,n 是折光指数。反射强度 $R(Q_z)$ 是 Q_z 的函数,可以由如下的两种方法测得:(1)改变入射中子束的波长 λ,保持入射角 θ 不变(此时需使用脉冲中子源);(2)选择固定的入射中子束的波长改变入射角(在反应堆中子源下)。

至于平面入射波,如果第一种介质是空气 $n_0 \approx 1$,那么所有的外部反射将在入射角小于临界入射角($\theta_0 = \theta_c$)的情况下就会发生,并且可以达到临界值 Q_c(该值是通过临界入射波长 λ_c 或临界入射角 θ_c 来定义的)。对于干净的重水 D_2O 的平面来说(见图 4.11(a)),如果 $\theta < \theta_c$(也就是在临界值 Q_c 以下)就会发生全反射,此时 $R(Q) \equiv 1$;而在临界值 Q_c 以上时,反射率是随着 Q^{-4} 的上升而急剧下降的,$R(Q) = 1$ 的区域通常用来确定散射因子的尺度,对于水亚相的表面活性剂单层而言(见图 4.11(b)),如果测量角度 $\theta < 1.5°$ 的话,一般都可以达到临界值 Q_c[22]。在穿过吸附层之后,入射的中子束将一部分发生折射,另一部分发生反射,而反射则是存在于界面膜的两侧的,所以,在两束反射波之间就会发生干涉现象,其结果是在 $R(Q)$ 的边缘出现"条纹",而最小值 Q_{min} 则与层厚度 τ 有关,$Q_{min} \approx 2\pi/\tau$(见图 4.11(b))。

分析表面活性剂单层或多层体系内的镜面反射率可以通过如下的比较进

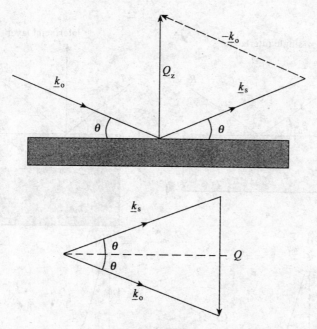

k_o 和 k_s 分别为入射及散射波向量；θ 为散射角

图 4.10 中子反射实验及散射向量 Q 的几何示意图

行，即比较由精确的光矩阵法计算的值和动力学近似法（Born 法）得到的值，而 Born 法则是源于经典的散射理论。

在第一种方法中定义了每一层的特征矩阵，通过 Fresnel 反射系数和相因数该特征矩阵将相邻层间的电矢量联系起来，其中 Fresnel 反射系数包含折光指数及反射角[23]。文献中对于该方法的详细报道也比较多[21,24]。该方法主要包括一系列的表面活性剂层，每一层都有其自己的散射长度密度 ρ 和厚度 τ，该值也包括了连续的两层之间的界面粗糙度。实际上，大部分的表面都不同程度地受到了界面粗糙度的影响，从而发生了镜面反射[25,26]，特别是液体表面甚至会出现因为受热而引发的表面张力波。把计算值和测量值加以比较，就会发现在每一层上 ρ 和 τ 都有不同程度的变化，直到二者的相容达到最佳（由最小二乘法迭代得到）。于是就可以相应地得到二级参数，如单分子的面积或者是表面覆盖度（如下所述）。

考虑到表面活性剂单层厚度 d 和折光指数 n_1 的简单性，我们引入体相折光指数 n_0 和 n_2（见图 4.11（b））。如果表面活性剂的溶液组成是无反射的水（NRW），其组成是摩尔分率为 8% 的 D_2O 加入到普通水中，此时 $\rho=0$ 而 $n=1$，所以，对于 NRW 来说就没有散射因素，所有的散射均是由表面活性剂单层引起

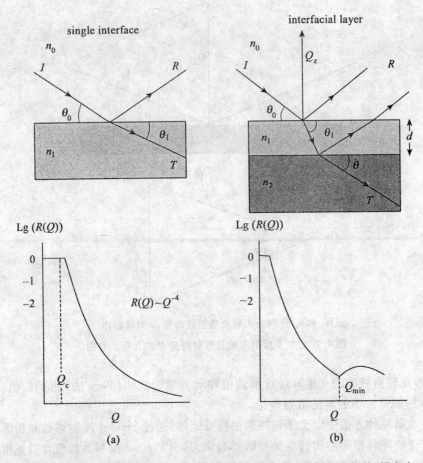

图 4.11 (a)存在明显界面的两相区的镜面反射,(b)处在两相之间的,厚度为 d 的薄层界面上的反射。I,R,T 分别代表入射、反射和折射中子束。其他的符号意义同文中的定义。同时还给出了在这两种情况下 $R(Q)$ 的示意图[22]。

的,如果把所有自表面活性剂单层所得的数据整理为一个模型的话,那么表面过剩 Γ 就可以计算如下:

$$A_s = \frac{\sum b_i}{\rho(z)\tau} = \frac{1}{\Gamma N_a} \tag{4.3.3}$$

式中,$\sum b_i$ 是单分子散射长度的总和,N_a 是阿佛加德罗常数,$\rho(z)$ 和 τ 是由上面提到的方法所确定的最佳散射密度和厚度。而 τ 的值是与所选择的模型有关的,并且随所选择用来描述正交于表面的散射密度的散射分布函数变化。然而,依照 Simister 等[29]的研究结果,能得到一个较好的拟合结果的 $\rho(z)$ 值的设定,也可精确补偿 τ 的变化,这样使得 A_s 不随 τ 的变化而变化。

第二种方法是动力学近似法(也称为 Born 近似法)[30]。按照这种方法,反射率 $R(Q_z)$ 与和界面正交的散射密度 $\rho(z)$ 有关:

$$R(Q_z) = \frac{16\pi^2}{Q_z^2} | \hat{\rho}(Q_z) |^2 \quad (4.3.4)$$

式中,$\hat{\rho}(Q_z)$ 是关于 $\rho(z)$ 的一维傅里叶变换。

$$\hat{\rho}(Q_z) = \int_{-\infty}^{+\infty} \exp(-iQ_z z)\rho(z)\mathrm{d}z \quad (4.3.5)$$

关于动力学近似法研究薄膜吸附的进展详情可以参见文献[30,31]。对于表面活性剂单层在空气-溶液界面上的吸附,人们研究的重点除了吸附膜的厚度、表面覆盖度之外,还包括表面活性剂链、头基与水的相对位置,以及相对于界面的正态分布的宽度。这些结构特征可以通过补充分析的 $R(Q)$ 曲线获得,$R(Q)$ 则由 H/D 同位素标记的表面活性剂检测得到。这就是部分结构因子分析法(PSF),在进行动力学近似的时候是非常有效的。通过标记方案,总的散射可以由多个部分结构因子来表示,而这些结构因子是用来描述界面上的各种组分的[30,32]。接下来就是对于简单的两组分体系(表面活性剂(A)和溶剂(S))的分析。基于上面的说明,散射密度 $\rho(z)$ 可以表示为

$$\rho(z) = N_S(z)b_S + N_A(z)b_A \quad (4.4.6)$$

式中,$N_A(z)$ 和 $N_S(z)$ 分别是溶液和溶剂的分子数密度,而 b 则是散射长度。联合(4.3.4)式和(4.3.6)式就可以得到

$$R(Q_z) = \frac{16\pi^2}{Q_z^2}\left[b_A^2 h_{AA} + b_S^2 h_{SS} + 2b_A b_S h_{AS}\right] \quad (4.3.7)$$

式中,的 h_{ii} 和 h_{ij} 是部分结构因子(PSF),h_{ii} 是包含各组分分布信息的同项,并且 h_{ii} 是关于 $N_i(z)$ 的一维傅里叶变换:

$$h_{ii}(Q_z) = |N_i(Q_z)|^2 \quad (4.3.8)$$

h_{ij} 两组分之间的交叉项,可以描述不同组分之间的相互位置。在上面给出的关于表面活性剂的溶液和溶剂的例子中,假设 A 和 S 在界面的分布分别是偶(关于中心对称)奇分布[33],则有下式:

$$h_{AS} = \pm (h_{AA}h_{SS})^{1/2}\sin(Q_z\delta_{AS}) \quad (4.3.9)$$

式中,δ 是表面活性剂和溶剂分布中心之间的距离差。分布有可能不是精确的是偶/奇分布,对该假设的偏离可能影响结果的准确性,而由该结果可得到 δ_{AS}。对该近似不适用的情况在此不做详细的讨论[34],但对于 AOT 以及它的衍生物一般不会发生该情况[35,36]。

原则上,$N_i(z)$可以由 PSF 的傅里叶变换得到,但实际上,我们又可以完全表达这一值的分析函数,该函数是由傅里叶变换而来,适合于实验数据。对于可溶的表面活性剂单层,该函数表明,Gaussian 可以更好地表示数密度图 $N_A(z)$[37]:

$$N_A(z) = N_{A0}\exp\left(\frac{-4z^2}{\sigma_A^2}\right) \quad (4.3.10)$$

式中,N_{A0}是数密度的最大值,σ_A是最大数密度 $1/e$ 时分布宽度。表面活性剂单层的吸附量 Γ_m 与 N_{A0} 可以由下式关联:

$$\Gamma_m = \frac{1}{A_s} = \frac{\sigma_A N_{A0} \pi^{1/2}}{2} \quad (4.3.11)$$

式中,A_s 为每个分子的面积而对于界面上溶剂的分布。常用的分析形式是通过双曲正切给出的:

$$N_S = N_{S0}\left[\frac{1}{2} + \frac{1}{2}\tanh\left(\frac{z}{\zeta}\right)\right] \quad (4.3.12)$$

式中,N_{S0}是水在体相溶液中的数密度,ζ 是宽度参数。于是由(4.44)式和(4.12)式确定的分布的 PSF 就是

$$Q_z^2 h_{AA} = \frac{\pi \sigma_A^2 N_{A0}^2 Q_z^2}{4}\exp\left(-\frac{Q_z^2 \sigma_A^2}{8}\right) \quad (4.13)$$

$$Q_z^2 h_{SS} = \frac{N_{S0}^2 \zeta^2 \pi^2 Q_z^2}{4}\cosh^2\left(\frac{\zeta \pi Q_z}{2}\right) \quad (4.14)$$

交叉 PSF 即 h_{as} 可以由上面直接关联 h_{AA} 和 h_{SS} 的(4.43)式得到,同时还可以确定关于 δ_{AS} 的函数:

$$Q_z^2 h_{AS} = \frac{\sigma_A N_{S0} N_{A0} \pi^{3/2} \zeta Q_z^2}{4}\exp\left(-\frac{Q_z^2 \sigma_A^2}{16}\right)\cosh\left(\frac{\zeta \pi Q_z}{2}\right)\sin\pi\delta_{AS} \quad (4.3.15)$$

由动力学近似得到的结构参数(不包括 δ_{AS})与我们假设的分布的形状是没有关系的。溶质的结构参数(由表面覆盖度 Γ_m 表示)是与关于 $N_A(z)$ 的任何假设都没有关系的[30]。在水的表面上,对于具有 Gaussian 分布的表面活性剂单层,与空气和 NRW 下的情况相比较,反射率可以由下面的式子给出[38,39]

$$\frac{Q_z^2 R(Q_z)}{16\pi^2} \approx (\Gamma_m N_a b_A)^2 \exp(-Q_z^2 \sigma_A^2) \quad (4.16)$$

式中,N_a 为阿佛加德罗常数。于是以 $\ln Q_z^2 R(Q_z)$ 对 Q_z^2 的曲线外推到 $Q_z = 0$ 就是与 Γ_m 的值无关的模型。

动力学近似法假设所有的散射都是单散射,也就是说,忽略样品内部的多散射效应。所以,该近似只在散射极为微弱的条件下才成立,即入射光的强度(I_0)

要远远大于散射光的强度(I),在实际应用反射的时候,该近似在 $Q \sim Q_c$ 的区域附近不成立,因为此时 $I_s \sim I_0$,不能认为散射是极微弱的。所以,只有在远离全反射区域的时候,才可以利用该近似解释反射数据。因而(4.38)式只是近似的并且在 Q 值极低的时候根本不适用。为了利用全部实验范围内的数据,并且为了解释在 $Q_z \sim Q_c$ 附近,该近似不适用的原因,每一个 $R(Q_z)$ 的计算值在与检测值进行比较前必须要加以校正[28,40]。

目前,一篇优秀的关于中子反射仪应用的综述已经发表[41]。

参考文献(略)

1. Surfactant chemistry and general phase behaviour

1.1 SURFACTANTS IN COLLOIDAL SYSTEMS

The term *colloid* (which means "glue" in Greek) was first introduced in 1861 by Thomas Graham to describe the "pseudosolutions" in aqueous systems of silver chloride, sulfur, and Prussian blue which were prepared by Francesco Selmi in the mid-nineteenth century [1]. Such systems were characterised by a lack of sedimentation under the influence of gravity, as well as low diffusion rates. Graham thus deduced that the colloidal size range is approximately 1 μm down to 1 nm (i.e., 10^{-6}- 10^{-9} m). This characteristic still holds today and colloids are generally described as systems consisting of one substance finely dispersed in another. These substances are referred to as the dispersed phase and dispersion medium (or continuous phase) respectively, and can be a solid, a liquid, or a gas. Such combinations together with large surface areas associated with the characteristic size of colloidal particles give rise to a large variety of systems, practical applications and interfacial phenomena.

Amongst these systems, the most common and ancient class is probably the *lyophobic* ("liquid-hating") colloids, composed of insoluble or immiscible components. They can be traced back to the 1850's when Michael Faraday prepared colloidal gold sols, which involve solid particles in water [2]. More commonly encountered examples of lyophobic colloids are milk (liquid fat dispersed as fine drops in an aqueous phase), smoke (solid particles dispersed in air), fog (small liquid droplets dispersed in air), paints (small solid particles dispersed in liquid), jelly (large protein molecules dispersed in water), and bone (small particles of calcium phosphate dispersed in a solid matrix of collagen). A second and more recent class includes the *lyophilic* ("liquid-loving") colloids, which are solutions that form spontaneously and are thermodynamically stable. These systems consist of solute molecules that are polymers (i. e., of much larger size than the solvent molecules), and as such form a large and distinct area of research (polymer science).

Another major group of colloidal systems, also classified as lyophilic, is that of the so-called *association colloids*. These are aggregates of amphiphilic (both "oil and water-loving") molecules that associate in a dynamic and thermodynamically driven process that may be simultaneously a molecular solution and a true colloidal system. Such molecules are commonly termed "surfactants", a contraction of the term *surface-active agents*. As will be introduced below and described in more detail in Chapter 2, surfactants are an important and versatile class of chemicals. Due to their dual nature, they are associated with many useful interfacial phenomena, e. g., wetting, and as such are found in many diverse industrial products and processes.

1.2 CHARACTERISTIC FEATURES OF SURFACTANTS

Surface-active agents are organic molecules that, when dissolved in a solvent at low concentration, have the ability to adsorb (or locate) at interfaces, thereby altering significantly the physical properties of those interfaces. The term "interface" is commonly employed here to describe the boundary in liquid/liquid, solid/liquid and gas/liquid systems, although in the latter case the term "surface" can also be used. This adsorption behaviour can be attributed to the solvent nature and to a chemical structure for surfactants that combine both a polar and a non-polar (amphiphilic) group into a single molecule. To accommodate for their dual nature, amphiphiles therefore "sit" at interfaces so that their lyophobic moiety keeps away from strong solvent interactions while the lyophilic part remains in solution. Since water is the most common solvent, and is the liquid of most academic and industrial interest, amphiphiles will be described with regard to their "hydrophilic" and "hydrophobic" moieties, or "head" and "tail" respectively.

Adsorption is associated with significant energetic changes since the free energy of a surfactant molecule located at the interface is lower than that of a molecule solubilised in either bulk phase. Accumulation of amphiphiles at the interface (liquid/liquid or gas/liquid) is therefore a spontaneous process and results in a decrease of the interfacial (surface) tension. However, such a definition applies to many substances: medium- or long-chain alcohols are surface active (e.g., n-hexanol, dodecanol) but these are not considered as surfactants. True surfactants are distinguished by an ability to form oriented monolayers at the interface (here air/water or oil/water) and, most importantly, self-assembly structures (micelles, vesicles) in bulk phases. They also stand out from the more general class of surface-active agents owing to emulsification, dispersion, wetting, foaming or detergency properties.

Both adsorption and aggregation phenomena result from the hydrophobic effect [3]; i.e., the expulsion of surfactant tails from water. Basically this originates from water-water intermolecular interactions being stronger than those between water-tail. Finally another characteristic of surfactants, when their aqueous concentration exceeds approximately 40%, is an ability to form liquid crystalline phases (or lyotropic mesophases). These systems consist of extended aggregation of surfactant molecules into large organised structures.

Owing to such a versatile phase behaviour and diversity in colloidal structures, surfactants find application in many industrial processes, essentially where high surface areas, modification of the interfacial activity or stability of colloidal systems are required. The variety of surfactants and the synergism offered by mixed-surfactant systems [4] also explains the ever-growing interest in fundamental studies and practical applications. Listing the various physical properties and associated uses of surfactants is beyond the scope of this chapter. However, a few relevant examples are presented in the following section, giving an idea of their widespread industrial use.

1.3 CLASSIFICATION AND APPLICATIONS OF SURFACTANTS

1.3.1 Types of surfactants

Numerous variations are possible within the structure of both the head and tail group of surfactants. The head group can be charged or neutral, small and compact in size, or a polymeric chain. The tail group is usually a single or double, straight or branched hydrocarbon chain, but may also be a fluorocarbon, or a siloxane, or contain aromatic group(s). Commonly encountered hydrophilic and hydrophobic groups are listed in Tables 1.1 and 1.2 respectively.

Since the hydrophilic part normally achieves its solubility either by ionic interactions or by hydrogen bonding, the simplest classification is based on surfactant head group type, with further subgroups according to the nature of the lyophobic moiety. Four basic classes therefore emerge as:
- the anionics and cationics, which dissociate in water into two oppositely charged species (the surfactant ion and its counterion),
- the non-ionics, which include a highly polar (non charged) moiety, such as polyoxyethylene (—OCH_2CH_2O—) or polyol groups,
- the zwitterionics (or amphoterics), which combine both a positive and a negative group.

With the continuous search for improving surfactant properties, new structures have recently emerged that exhibit interesting synergistic interactions or enhanced surface and aggregation properties. These novel surfactants have attracted much interest and include the catanionics, bolaforms, gemini (or dimeric) surfactants, polymeric and polymerisable surfactants [5, 6]. Characteristics and typical examples are shown in Table 1.3. Another important driving force for this research is the need for enhanced surfactant biodegradability. In particular for personal care products and household detergents, regulations [7] require high biodegradability and non-toxicity of each component present in the formulation.

Surfactant Chemistry

Table 1.1 Common hydrophilic groups found in commercially available surfactants

Class	General structure
Sulfonate	$R-SO_3^- M^+$
Sulfate	$R-OSO_3^- M^+$
Carboxylate	$R-COO^- M^+$
Phosphate	$R-OPO_3^- M^+$
Ammonium	$R_x H_y N^+ X^-$ ($x = 1-3, y = 4-x$)
Quaternary ammonium	$R_4 N^+ X^-$
Betaines	$RN^+(CH_3)_2 CH_2 COO^-$
Sulfobetaines	$RN^+(CH_3)_2 CH_2 CH_2 SO_3^-$
Polyoxyethylene (POE)	$R-OCH_2 CH_2 (OCH_2 CH_2)_n O$
Polyols	Sucrose, sorbitan, glycerol, ethylene glycol, etc
Polypeptide	$R-NH-CHR-CO-NH-CHR'-CO\cdots\cdots CO_2 H$
Polyglycidyl	$R-(OCH_2 CH[CH_2 OH]CH_2)n\cdots\cdots OCH_2 CH[CH_2 OH]CH_2 OH$

Table 1.2 Common hydrophobic groups used in commercially available surfactants

Group	General structure	
Natural fatty acids	$CH_3(CH_2)_n$	$n = 12-18$
Petroleum paraffins	$CH_3(CH_2)_n CH_3$	$n = 8-20$
Olefins	$CH_3(CH_2)_n CH=CH_2$	$n = 7-17$
Alkylbenzenes	$CH_3(CH_2)_n CH_2$—⟨benzene⟩	$n = 6-10$, linear or branched
Alkylaromatics	$CH_3(CH_2)_n CH_3$ / R—⟨naphthalene⟩—R	$n = 1-2$ for water soluble, $n = 8$ or 9 for oil soluble surfactants

80

1. Surfactant chemistry and general phase behaviour

Table 1.2 continue

Group	General structure	
Alkylphenols	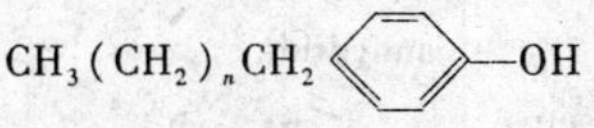 $CH_3(CH_2)_n CH_2$—⟨phenyl⟩—OH	$n = 6\text{-}10$, linear of branched

Polyoxypropylene	$CH_3CHCH_2O(CHCH_2)_n$
	    | |
	   X CH_3
		n = degree of oligomerisation, X = oligomerisation initiator

Fluorocarbons	$CF_3(CF_2)_n COOH$	$n = 4\text{-}8$, linear of branched, or H-terminated

Silicones	CH_3
	   |
	$CH_3O(SiO)_n CH_3$
	   |
	   CH_3	

Table 1.3 **Structural features and examples of new surfactant classes**

Classes	Structural characteristics	Example
Catanionic	Equimolar mixture of cationic and anionic surfactants (no inorganic counterion)	n-dodecyltrimethylammonium n-dodecyl sulfate (DTADS) $C_{12}H_{25}(CH_3)_3N^+ {}^-O_4SC_{12}H_{25}$
Bolaform	Two charged headgroups connected by a long linear polymethylene chain	Hexadecanediyl-1',1 6-bis(trimethyl ammonium bromide) $Br^-(CH_3)_3N^+-(CH_2)_{16}-N^+(CH_3)_3Br^-$
Gemini (or dimeric)	Two identical surfactants connected by a spacer close to or at the level of the headgroup	Propane-1,3-bis(dodecyldimethyl ammonium bromide) $C_3H_6\text{-}1,3\text{-bis}[(CH_3)_2N^+C_{12}H_{25}\,Br]$

81

Table 1.3 continue

Classes	Structural characteristics	Example
Polymeric	Polymer with surface active properties	Copolymer of isobutylene and succinic anhydride 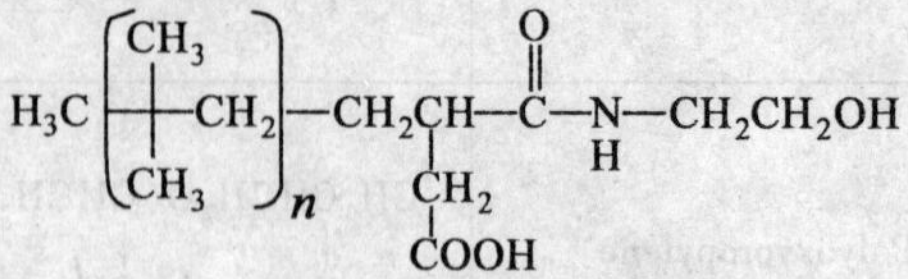 $H_3C{-}[{-}\underset{CH_3}{\overset{CH_3}{C}}{-}CH_2{-}]_n{-}CH_2CH{-}\underset{\underset{COOH}{CH_2}}{}{-}\overset{O}{\overset{\|}{C}}{-}\underset{H}{N}{-}CH_2CH_2OH$
Polymerisable	Surfactant that can undergo homo-polymerisation or copolymerisation with other components of the system	11-(acryloyloxy)undecyltrimethyl ammonium bromide 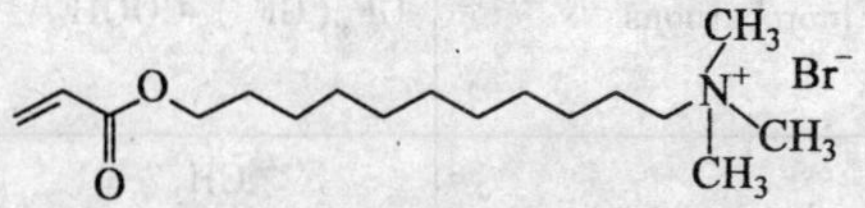

A typical example of a double-chain surfactant is sodium bis(2-ethylhexyl)sulfosuccinate, often referred to by its American Cyanamid trade name Aerosol-OT, or AOT. Its chemical structure is illustrated in Figure 1.1, along with other typical double-chain compounds within the four basic surfactant classes.

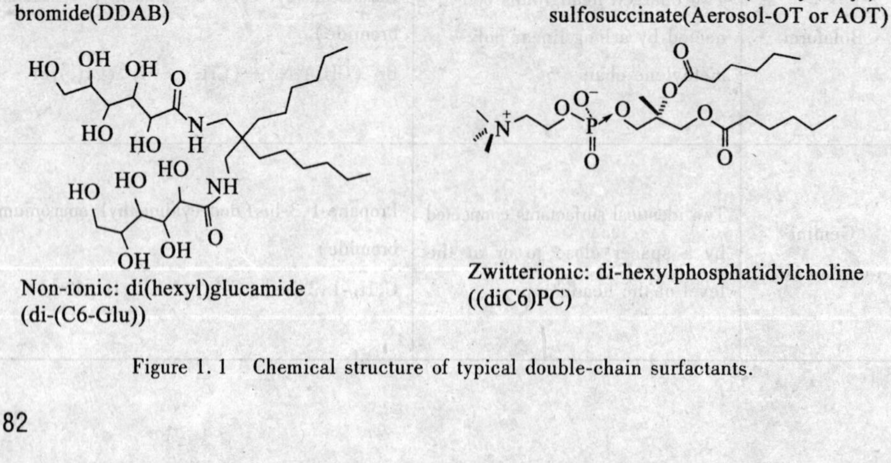

Cationic: n-didodecyldimethylammonium bromide(DDAB)

Anionic: Sodium bis(2-ethylhexyl) sulfosuccinate(Aerosol-OT or AOT)

Non-ionic: di(hexyl)glucamide (di-(C6-Glu))

Zwitterionic: di-hexylphosphatidylcholine ((diC6)PC)

Figure 1.1 Chemical structure of typical double-chain surfactants.

1.3.2 Surfactant uses and development

Surfactants may be from natural or synthetic sources. The first category includes naturally occurring amphiphiles such as the lipids, which are surfactants based on glycerol and are vital components of the cell membrane. Also in this group are the so-called "soaps", the first recognised surfactants [8]. These can be traced back to Egyptian times; by combining animal and vegetable oils with alkaline salts a soap-like material was formed, and this was used for treating skin diseases, as well as for washing. Soaps remained the only source of natural detergents from the seventh century till the early twentieth century, with gradually more varieties becoming available for shaving and shampooing, as well as bathing and laundering. In 1916, in response to a World War I-related shortage of fats for making soap, the first synthetic detergent was developed in Germany. Known today simply as detergents, synthetic detergents are washing and cleaning products obtained from a variety of raw materials.

Nowadays, synthetic surfactants are essential components in many industrial processes and formulations [9-11]. Depending on the precise chemical nature of the product, the properties of, for example emulsification, detergency and foaming may be exhibited in varying degree. The number and arrangement of the hydrocarbon groups together with the nature and position of the hydrophilic groups combine to determine the surface-active properties of the molecule. For example C12 to C20 is generally regarded as the range covering optimum detergency, whilst wetting and foaming are best achieved with shorter chain lengths. Structure-performance relationships and chemical compatibility are therefore key elements in surfactant-based formulations, so that much research is devoted to this area.

Amongst the different classes of surfactants, anionics are often used in greater volume than any other types, mainly because of the ease and low cost of manufacture. They contain negatively charged head group, e. g. , carboxylates ($-CO_2^-$), used in soaps, sulfate ($-OSO_3^-$), and sulfonates ($-SO_3^-$) groups. Their main applications are in detergency, personal care products, emulsifiers and soaps.

Cationics have positively charged head groups—e.g., trimethylammonium ion ($-N(CH_3)_3^+$)— and are mainly involved in applications related to their absorption at surfaces. These are generally negatively charged (e.g., metal, plastics, minerals, fibres, hairs and cell membranes) so that they can be modified upon treatment with cationic surfactants. They are therefore used as anticorrosion and antistatic agents, flotation collectors, fabric softeners, hair conditioners and bactericides.

Non-ionics contain groups with a strong affinity for water due to strong dipole-dipole interactions arising from hydrogen bonding, e.g., ethoxylates ($-(OCH_2CH_2)_mOH$). One advantage over ionics is that the length of both the hydrophilic and hydrophobic groups can be varied to obtain maximum efficiency in use. They find applications in low temperature detergents and emulsifiers.

Zwitterionics constitute the smallest surfactant class due to their high cost of manufacture. They are characterised by excellent dermatological properties and skin compatibility. Because of their low eye and skin irritation, common uses are in shampoos and cosmetics.

REFERENCES

1. Evans, D. F.; Wennerström, H. *The Colloidal Domain*, Wiley-VCH, 1999, New York.
2. Faraday, M., Phil. Trans. Royal. Soc., 1857, 147, 145.
3. Tanford, C. *The Hydrophobic Effect: formation of micelles and biological membranes*, John Wiley & Sons, 1978, USA.
4. Ogino, K.; Abe, M., Eds. *Mixed Surfactants Systems*, Marcel Dekker, 1993, New York.
5. Robb, I. D. *Specialist Surfactants*, Blackie Academic & Professional, 1997, London.
6. Holmberg, K. Ed. *Novel Surfactants*, Marcel Dekker, 1998, New York.

7. Hollis, G. Ed. '*Surfactants UK*', Tergo-Data, 1976.
8. The Soap and Detergent Association home page, http://www.sdahq.org/.
9. Karsa, D. R.; Goode, J. M.; Donnelly, P. J. Eds. '*Surfactants Applications Directory*', Blackie & Son, 1991, London.
10. Dickinson, E. in '*An Introduction to Food Colloids*', Oxford University Press, 1992, Oxford.
11. Solans, C.; Kunieda, H. Eds. '*Industrial Applications of Microemulsions*' Marcel Dekker, 1997, New York.

2. Aggregation and adsorption at interfaces

Surfactants, literally, are active at a surface and that includes any of the liquid/liquid, liquid/gas or liquid/solid systems, so that the subject is quite broad. In this chapter particular emphasis is placed on adsorption and aggregation phenomena in aqueous systems. For a more thorough account of the theoretical background of surfactancy, the reader is referred to specific textbooks and monographs keyed throughout this chapter.

2.1 ADSORPTION OF SURFACTANTS AT INTERFACES

2.1.1 Surface tension and surface activity

Due to the different environment of molecules located at an interface compared to those from either bulk phase, an interface is associated with a surface free energy. At the air-water surface for example, water molecules are subjected to unequal short-range attraction forces and, thus, undergo a net inward pull to the bulk phase. Minimisation of the contact area with the gas phase is therefore a spontaneous process, explaining why drops and bubbles are round. The surface free energy per unit area, defined as the surface tension (γ_o), is then the minimum amount of work (W_{min}) required to create new unit area of that interface (ΔA), so $W_{min} = \gamma_o \times \Delta A$. Another, but less intuitive, definition of surface tension is given as the force acting normal to the liquid-gas interface per unit length of the resulting thin film on the surface.

A surface-active agent is therefore a substance that at low concentrations adsorbs thereby changing the amount of work required to expand that interface. In particular surfactants can significantly reduce interfacial tension due to their dual chemical nature as introduced in Chapter 1. Considering the air-water boundary, the force driving adsorption is unfavourable hydrophobic interactions within the bulk phase. There, water molecules strongly interact with one another through van der waals forces and hydrogen bonding, so the presence of amphiphilic molecules dissolved in the bulk phase, through their hyrocarbon group, causes distortion of this solvent structure increasing the free energy of the system. This is known as the hydrophobic effect [1]. Less work is required to bring a surfactant molecule to the surface than a water molecule, so that migration of the surfactant to the surface is a spontaneous and favourable process. At the gas-liquid interface, the result is the creation of new unit area of surface and the formation of an *oriented surfactant monolayer* with the hydrophobic tails pointing out of, and the head group inside, the water phase. The balance against the tendency of the surface to contract under normal surface tension forces causes an increase in the surface (or expanding) pressure π, and therefore a decrease in surface tension of the solution γ. The surface pressure is defined as $\pi = \gamma_0 - \gamma$, where γ_0 is the surface tension of a clean air-water surface.

Depending on the surfactant molecular structure, adsorption takes place over various concentration ranges and rates, but typically, above a well-defined concentration—the critical micelle concentration (CMC)—micellisation or aggregation takes place. At the CMC, the interface is at (near) maximum coverage and to minimise further free energy, molecules begin to aggregate in the bulk phase. Above the CMC, the system then consists of an adsorbed monomolecular layer, free monomers and micellised surfactant in the bulk, with all these three states in equilibrium. The structure and formation of micelles will be briefly described in Section 2.3. Below the CMC, adsorption is a dynamic equilibrium with surfactant molecules arriving at, and leaving, the surface. at equal rate. Nevertheless, a time-averaged value for the surface concentration can be defined and quantified either directly or indirectly using thermodynamic equations (see Section 2.1.2).

Dynamic surface tension—as opposed to the equilibrium quantity—is an important property of surfactant systems as it governs many important industrial and biological applications [2-5]. Examples are printing and coating processes where an equilibrium surface tension is never attained, and a new area of interface is continuously formed. In any surfactant solution, the equilibrium surface tension is not achieved instantaneously and surfactant molecules must first diffuse from the bulk to the surface, then adsorb, whilst also achieving the correct orientation. Therefore, a freshly formed interface of a surfactant solution has a surface tension very close to that of the solvent, and this dynamic surface tension will then decay over a certain period of time to the equilibrium value. This relaxation can range from milliseconds to days depending on the surfactant type and concentration. In order to control this dynamic behaviour, it is necessary to understand the main processes governing transport of surfactant molecules from the bulk to the interface. This area of research therefore attracts much attention and recent developments can be found in references [6-8]. However, in the present chapter equilibrium surface tension will always be considered.

2.1.2 Surface excess and thermodynamics of adsorption

Following on the formation of an oriented surfactant monolayer, a fundamental associated physical quantity is the *surface excess*. This is defined as the concentration of surfactant molecules in a surface plane, relative to that at a similar plane in the bulk. A common thermodynamic treatment of the variation of surface tension with composition has been derived by Gibbs [9].

An important approximation associated with this Gibbs adsorption equation is the "exact" location of the interface. Consider a surfactant aqueous phase α in equilibrium with vapour β. The interface is a region of indeterminate thickness τ across which the properties of the system vary from values specific to phase α to those characteristic of β. Since properties within this real interface cannot be well defined, a convenient assumption is to consider a mathematical plane, with zero thickness, so that the properties of α and β apply right up to that dividing plane positioned at some specific value X. Figure 2.1 illustrates this ideal system.

In the definition of the Gibbs dividing surface X is arbitrarily chosen so that the surface excess adsorption of the solvent is zero. Then the surface excess concentration of component i is given by

$$\Gamma_i^\sigma = \frac{n_i^\sigma}{A} \qquad (2.1.1)$$

2. Aggregation and adsorption at interfaces

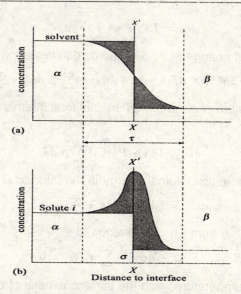

Figure 2.1 In the Gibbs approach to defining the surface excess concentration Γ, the Gibbs dividing surface is defined as the plane in which the solvent excess concentration becomes zero (the shaded area is equal on each side of the plane) as in (a). The surface excess of component i will then be the difference in the concentrations of that component on either side of that plane (the shaded area) (b).

where A is the interfacial area. The term n_i^σ is the amount of component i in the surface phase σ over and above that which would have been in the phase σ if the bulk phases α and β had extended to the surface XX', without any change of composition. Γ_i^σ may be positive or negative, and its magnitude clearly depends on the location of XX'.

Now consider the internal energy U of the total system consisting of the bulk phases α and β:

$$U = U^\alpha + U^\beta + U^\sigma$$

$$U^\alpha = TS^\alpha - PV^\alpha + \sum_i \mu_i n_i^\alpha \qquad (2.1.2)$$

$$U^\beta = TS^\beta - PV^\beta + \sum_i \mu_i n_i^\beta$$

The corresponding expression for the thermodynamic energy of the interfacial region σ is

$$U^\sigma = TS^\sigma + \gamma A + \sum_i \mu_i n_i^\sigma \tag{2.1.3}$$

For any infinitesimal change in T, S, A, μ, n, differentiation of Eq. 2.1.3 gives

$$dU^\sigma = TdS^\sigma + S^\sigma dT + \gamma dA + Ad\gamma + \sum_i \mu_i dn_i^\sigma + \sum_i n_i^\sigma d\mu_i \tag{2.1.4}$$

For a small, reversible change the differential total internal energy in any bulk phase is

$$dU = TdS - PdV + \sum_i \mu_i dn_i \tag{2.1.5}$$

similarly for the differential internal energy in the interfacial region:

$$dU^\sigma = TdS^\sigma + \gamma dA + \sum_i \mu_i dn_i^\sigma \tag{2.1.6}$$

subtracting Eq. 2.1.6 from Eq. 2.1.4 leads to

$$S^\sigma dT + Ad\gamma + \sum_i n_i^\sigma d\mu_i = 0 \tag{2.1.7}$$

Then at constant temperature, with the surface excess of component i, τ_i^σ, as defined in Eq. 2.1.1, the general form of the Gibbs equation is

$$d\gamma = - \sum_i \Gamma_i^\sigma d\mu_i \tag{2.1.8}$$

For a simple system consisting of a solvent and a solute, denoted by the subscripts 1 and 2 respectively, then Eq. 2.1.8 reduces to

$$d\gamma = - \Gamma_1^\sigma d\mu_1 - \Gamma_2^\sigma d\mu_2 \tag{2.1.9}$$

Considering the choice of the Gibbs dividing surface position, i.e., so that $\tau_1^\sigma = 0$, then Eq. 2.1.9 simplifies to

$$d\gamma = - \Gamma_2^\sigma d\mu_2 \tag{2.1.10}$$

where Γ_2^σ is the solute surface excess concentration.
The chemical potential is given by

$$\mu_i = \mu_i^0 + RT \ln a_i \quad \text{so} \quad d\mu_i = \text{cste} + RT d \ln a_i \tag{2.1.11}$$

where μ_i^0 is the standard chemical potential of component i at 1 Atm and 298K. Therefore applying to Eq. 2.1.10 gives the common form of the Gibbs equation for non-dissociating materials (e.g., non-ionic surfactants):

$$d\gamma = - \Gamma_2^\sigma RT d \ln a_2 \tag{2.1.12}$$

or

$$\Gamma_2^\sigma = - \frac{1}{RT} \frac{d\gamma}{d \ln a_2} \tag{2.1.13}$$

For dissociating solutes, such as ionic surfactants of the form $R^- M^+$ and assuming ideal behaviour below the CMC, Eq. 2.1.12 becomes

$$d\gamma = - \Gamma_R^\sigma d\mu_R - \Gamma_M^\sigma d\mu_M \tag{2.1.14}$$

2. Aggregation and adsorption at interfaces

If no electrolyte is added, electroneutrality of the interface requires that $\Gamma_R^\sigma = \Gamma_M^\sigma$. Using the mean ionic activities so that $a_2 = (a_R a_M)^{1/2}$ and substituting in Eq. 2.1.14 gives the Gibbs equation for 1:1 dissociating compounds:

$$\Gamma_2^\sigma = -\frac{1}{2RT}\frac{d\gamma}{d\ln a_2} \qquad (2.1.15)$$

If swamping electrolyte is introduced (i.e., sufficient salt to make electrostatic effects unimportant) and the same gegenion M^+ as the surfactant is present, then the activity of M^+ is constant and the pre-factor becomes unity, so that Eq. 2.1.13 is appropriate.

For materials that are strongly adsorbed at an interface such as surfactants, a dramatic reduction in interfacial (surface) tension is observed with small changes in bulk phase concentration. The practical applicability of this relationship is that the relative adsorption of a material at an interface, its surface activity, can be determined from measurement of the interfacial tension as a function of solute concentration. Note that in Eq. 2.1.13 and 2.1.15, for dilute surfactant systems, the concentration can be substituted for activity without loss of generality.

Figure 2.2 shows a typical decay of surface tension of water on increase in surfactant concentration, and how the Gibbs equation (Eq. 2.1.13 or 2.1.15) is used to quantify adsorption at the surface. At low concentrations a gradual decay in surface tension is observed (from the surface tension of pure water i.e., 72.5 mN·m^{-1} at 25°C) corresponding to an increase in the surface excess of component 2 (region A to B). Then at concentrations close to the CMC, the adsorption tends to a limiting value so the surface tension curve may appear to be essentially linear (region B to C). However, in practice, for most surfactants in the pre-CMC region the γ-ln c is curved so that the local tangent $-d\gamma/d\ln c$ is proportional to Γ_2^σ via Eq. 2.1.13 or Eq. 2.1.15. For single-chain, pure surfactants typical values for Γ_2^σ at the CMC are in the range $2-4 \times 10^{-6}$ mol·m^{-2}, with the associated limiting molecular areas being from 0.4-0.6 nm^2.

The value for the Gibbs pre-factor in the case of ionic surfactants has recently been a matter of discussion (e.g., refs. 10-13). Of particular concern is the question whether, in the case of ionics, complete dissociation occurs giving rise to a pre-factor of 2, or a depletion layer in the sub-surface

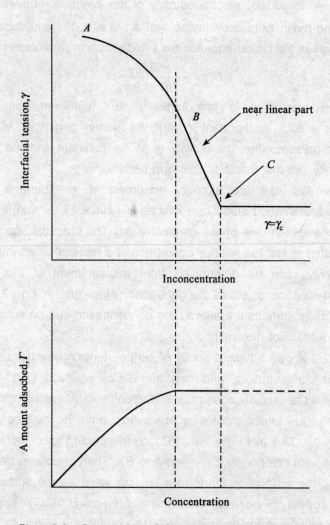

Figure 2.2 Determination of the interfacial adsorption isotherm from surface tension measurement and the Gibbs adsorption equation.

could be present so that a somewhat lower pre-factor could be expected. Recent detailed experiments combining tensiometry and neutron reflectivity, which enables direct measurement of the surface excess (as detailed in Chapter 4), have confirmed the use of a pre-factor of 2 in the case of ionics [14].

Although the Gibbs equation is the most commonly used mathematical relation for adsorption at liquid-liquid and liquid-gas interfaces, other adsorption isotherms have been proposed such as the Langmuir [15], the Szyszkowski [16] and the Frumkin [17] equations. The Gibbs equation itself has been simplified by Guggenheim and Adam with the choice of a different dividing plane and where the interfacial region is considered as a separate bulk phase (of finite volume) [18].

2.1.3 Efficiency and effectiveness of surfactant adsorption

The performance of a surfactant in lowering the surface tension of a solution can be discussed in terms of (1) the concentration required to produce a given surface tension reduction and (2) the maximum reduction in surface tension that can be obtained regardless of the concentration. These are referred to as the surfactant *efficiency* and *effectiveness* respectively.

A good measure of the *surfactant adsorption efficiency* is the concentration of surfactant required to produce a 20 mN · m^{-1} reduction in surface tension. At this value the surfactant concentration is close to the minimum concentration needed to produce maximum adsorption at the interface. This is confirmed by the Frumkin adsorption Eq. 2.1.16, which relates the reduction in surface tension (or surface pressure π) and surface excess concentration.

$$\gamma_0 - \gamma = \pi = -2.303 RT \Gamma_m \lg\left(1 - \frac{\Gamma_1}{\Gamma_m}\right) \quad (2.1.16)$$

The maximum surface excess generally lies in the range 1-4.4 × 10^{-10} mol · cm^{-2} [19]: solving Eq. 2.1.16 indicates that when the surface tension has been reduced by 20 mN · m^{-1}, at 25℃, the surface is 84%-99.9% saturated. The negative logarithm of such concentration, pC_{20}, is then a useful quantity since it can be related to the free energy change $\Delta G°$ involved in the transfer of a surfactant molecule from the interior of the bulk liquid phase to the interface. The surfactant adsorption efficiency thus relates to the structural groups in the molecule via the standard free energy change of the individual groups (i.e., free energies of transfer of methylene, terminal methyl, and head groups). In particular, for a given homologous series of straight-chain surfactants in water, $CH_3(CH_2)_n$—M, where M is the hydrophilic head group and n is the number of methylene units in the chain, and when the systems are at $\pi = 20$ mN·m^{-1}, the standard free energy of adsorption is

$$\Delta G° = n\Delta G°(-CH_2-) + \Delta G°(M) + \Delta G°(CH_3-) \quad (2.1.17)$$

Then the adsorption efficiency is directly related to the length of the hydrophobic chain (the hydrophilic group remains the same), viz.

$$-\lg(C)_{20} = pC_{20} = n\left[\frac{-\Delta G°(-CH_2-)}{2.303RT}\right] + \text{constant} \quad (2.1.18)$$

Where $\Delta G°(M)$ is considered as a constant and it is assumed that Γ_m does not differ significantly with increasing chain length, and that activity coefficients are unity. The efficiency factor pC_{20} therefore increases linearly with the number of carbon atoms in the hydrophobic chain. This is also described by Traube's rule [20] (Eq. 2.1.19).

$$\lg C_s = B - n\lg K_T \quad (2.1.19)$$

Where C_s is the surfactant concentration, B is a constant, n is the chain length within a homologous series and K_T is Traube's constant. For hydrocarbon straight chain surfactants K_T is usually around 3 [21] or by analogy to Eq. 2.1.18 is given by

$$\frac{C_n}{C_{n+1}} = K_T = \exp\left[\frac{-\Delta G°(-CH_2-)}{2RT}\right] \quad (2.1.20)$$

For compounds having a phenyl group in the hydrophobic chain it is equivalent to about three and one-half normal $-CH_2-$ groups.

The larger pC_{20} the more efficiently the surfactant is adsorbed at the interface and the more efficiently it reduces surface tension. The other main factors that contribute to an increase in surfactant efficiency are summarised below:

- A straight alkyl chain as the hydrophobic group, rather than a branched alkyl chain containing the same number of carbon atoms.
- A single hydrophilic group situated at the end of the hydrophobic group, rather than one (or more) at a central position.
- A non-ionic or zwitterionic hydrophilic group, rather than an ionic one.
- For ionic surfactants, a reduction in the effective charge by (a) use of a more tightly bound (less hydrated) counterion and (b) increase in ionic strength of the aqueous phase.

The choice of 20 mN · m^{-1} as a standard value of surface tension lowering for the definition of adsorption efficiency is convenient but somewhat arbitrary, and is not valid for systems where surfactants differ significantly in maximum surface excess or when the surface pressure is less than 20 mN · m^{-1}. Pitt et al. [22] circumvented this problem by defining $\Delta\gamma$ as half the surface pressure at the CMC.

The performance of a surfactant can also be discussed in terms of *effectiveness of adsorption*. This is usually defined as the maximum lowering of surface tension γ_{min} regardless of concentration, or as the surface excess concentration at surface saturation Γ_m since it represents the maximum adsorption. γ_{min}, and Γ_m, are controlled mainly by the critical micelle concentration, and for certain ionics by the solubility limit or Krafft temperature T_k, which will be described briefly in Section 2.2.1. The effectiveness of adsorption is an important factor in determining such properties as foaming, wetting, and emulsification, since Γ_m through the Gibbs adsorption equation gives a measure of the interfacial packing.

The efficiency and effectiveness of surfactants do not necessarily run parallel, and it is commonly observed—as shown by Rosen's extensive data listing [19]— that materials producing significant lowering of the surface tension at low concentrations (i.e., they are more efficient) have smaller Γ_m (i.e., they are less effective). In determining surfactant efficiency the role of the molecular structure is primarily thermodynamic, while its role in effectiveness is directly related to the relative size of the hydrophilic and hydrophobic portions of the adsorbing molecule. The area occupied by each molecule is determined either by the hydrophobic chain cross-sectional area, or the area required for closest packing of head groups, whichever is greater. Therefore, surfactant films can be tightly or loosely packed resulting in very different interfacial properties. For instance, straight chains and large head groups (relative to the cross tail section) favour close, effective packing, while branched, bulky, or multiple hydrophobic chains give rise to steric hindrance at the interface. On the other hand, within a series of single straight chain surfactants, increasing the hydrocarbon chain length from C8 to C20 will have little effect on adsorption effectiveness [19].

2.2 SURFACTANT SOLUBILITY

In aqueous solution, when all available interfaces are saturated, the overall energy reduction may continue through other mechanisms. Depending on the system composition, a surfactant molecule can play different roles in terms of aggregation (formation of micelles, liquid crystal phases, bilayers or vesicles, etc). The physical manifestation of one such mechanism is crystallisation or precipitation of surfactant from solution—that is, bulk-phase separation. While most common surfactants have a substantial solubility in water, this can change significantly with variations in hydrophobic tail length, head group nature, counterion valence, solution environment, and most importantly, temperature.

2.2.1 The Krafft temperature

As for most solutes in water, increasing temperature produces an increase in solubility. However, for ionic surfactants, which are initially insoluble, there is often a temperature at which the solubility suddenly increases very dramatically. This is known as the Krafft point or Krafft temperature, T_K, and is defined as the intersection of the solubility and the CMC curves, i.e., it is the temperature at which the solubility of the monomeric surfactant is equivalent to its CMC at the same temperature. This is illustrated in Figure 2.3. Below T_K, surfactant monomers only exist in equilibrium with the hydrated crystalline phase, and above T_K, micelles are formed providing much greater surfactant solubility.

The Krafft point of ionic surfactants is found to vary with counterion [23], alkyl chain length and chain structure. Knowledge of the Krafft temperature is crucial in many applications since below T_K the surfactant will clearly not perform efficiently; hence typical characteristics such as maximum surface tension lowering and micelle formation cannot be achieved. The development of surfactants with a lower Krafft point but still being very efficient at lowering surface tension (i.e., long chain compounds) is usually achieved by introducing chain branching, multiple bonds in the alkyl chain or bulkier hydrophilic groups thereby reducing intermolecular interactions that promote crystallisation.

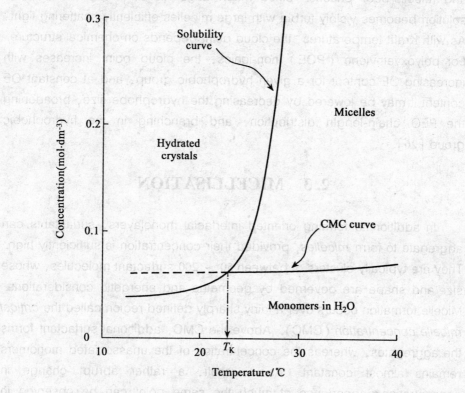

Figure 2.3 The Krafft temperature T_K is the point at which surfactant solubility equals the critical micelle concentration. Above T_K, surfactant molecules form a dispersed phase; below T_K, hydrated crystals are formed.

2.2.2 The Cloud point

For non-ionic surfactants, a common observation is that micellar solutions tend to become visibly turbid at a well-defined temperature. This is often referred to as the cloud point, above which the surfactant solution phase separates. Above the cloud point, the system consists of an almost micelle-free dilute solution at a concentration equal to its CMC at that temperature, and a surfactant-rich micellar phase. This separation is caused by a sharp increase in aggregation number and a decrease in intermicellar repulsions [24,25] that produces a difference in density of the micelle-rich and micelle-poor phases. Since much larger particles are formed, the solution becomes visibly turbid with large micelles efficiently scattering light. As with Krafft temperatures, the cloud point depends on chemical structure. For polyoxyethylene (POE) non-ionics, the cloud point increases with increasing OE content for a given hydrophobic group, and at constant OE content it may be lowered by decreasing the hydrophobe size, broadening the PEO chain-length distribution, and branching in the hydrophobic group [26].

2.3 MICELLISATION

In addition to forming oriented interfacial monolayers, surfactants can aggregate to form *micelles*, provided their concentration is sufficiently high. They are typically clusters of between 50 ~ 200 surfactant molecules, whose size and shape are governed by geometric and energetic considerations. Micelle formation occurs over a fairly sharply defined region called the *critical micelle concentration* (CMC). Above the CMC, additional surfactant forms the aggregates, whereas the concentration of the unassociated monomers remains almost constant. As a result, a rather abrupt change in concentration dependence at much the same point can be observed in common equilibrium or transport properties (Figure 2.4).

2.3.1 Thermodynamics of micellisation

Micelles are dynamic species, in that there is a constant, rapid interchange—typically on a microsecond timescale—of molecules between the aggregate and solution pseudo-phases. This constant formation-dissociation process relies on a subtle balance of interactions. These come from contacts between (1) hydrocarbon chain—water, (2) hydrocarbon—hydrocarbon chains, (3) head group—head group, and (4) from solvation of the head group. Therefore, the net free energy change upon micellisation, ΔG_m, can be written as

$$\Delta G_m = \Delta G(\text{HC}) + \Delta G(\text{contact}) + \Delta G(\text{packing}) + \Delta G(\text{HG})$$

(2.3.1)

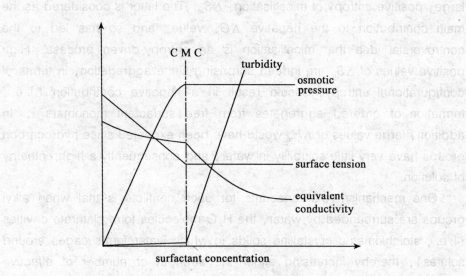

Figure 2.4 Schematic representation of the concentration dependence of some physical properties for solutions of a micelle-forming surfactant.

where
- $\Delta G(\text{HC})$ is the free energy associated with transferring hydrocarbon chains out of water and into the oil-like interior of the micelle.
- ΔG (contact) is a surface free energy attributed to solvent-hydrocarbon contacts in the micelle.
- $\Delta G(\text{packing})$ is a positive contribution associated with confining the hydrocarbon chain to the micelle core.
- ΔG (HG) is a positive contribution associated with head group interactions, including electrostatic as well as head group conformation effects.

Aggregation of surfactant molecules partly results from the tendency of the hydrophobic groups to minimise contacts with water by forming oily microdomains within the solvent. There, alkyl-alkyl interactions are maximised, while hydrophilic head groups remain surrounded by water.

The traditional picture of micelle formation thermodynamics is based on the Gibbs-Helmholtz equation ($\Delta G_m = \Delta H_m - T\Delta S_m$). At room temperature the process is characterised by a small, positive enthalpy, ΔH_m, and a large, positive entropy of micellisation, ΔS_m. The latter is considered as the main contribution to the negative ΔG_m value, and so has led to the controversial idea that micellisation is an entropy-driven process. High positive values of ΔS_m are indeed surprising since aggregation, in terms of configurational entropy, should result in a negative contribution (i.e., formation of ordered aggregates from free surfactant monomers). In addition, large values of ΔH_m would have been expected since hydrocarbon groups have very little solubility in water, and consequently a high enthalpy of solution.

One mechanism that accounts for such conflicts is that when alkyl groups are surrounded by water, the H_2O molecules form clathrate cavities (i.e., stoichiometric crystalline solids in which water forms cages around solutes), thereby increasing either the strength or number of effective hydrogen bonds [27]. Therefore, the predominant effect of the hydrocarbon molecule is to increase the degree of structure in the immediately surrounding water. This is one of the main features of the *hydrophobic effect*, a subject that was explored in detail by Tanford [1] to account for the very slight solubility of hydrocarbons in water. During the formation of micelles, the reverse process occurs: as lyophobic residues aggregate, the highly structured water around each chain collapses back to ordinary bulk water thereby accounting for the apparent large overall gain in entropy, ΔS_m. This water-structure effect was also invoked by other researchers [28, 29].

Such an interpretation, however, has been strongly challenged by more recent studies of aqueous systems at high temperatures (up to 166°C) and micellisation in hydrazine solutions [30]. In these systems water loses most of its peculiar structural properties and the formation of structured water around lyophobic species is no longer possible.

The mechanism of micelle formation from surfactant monomers, S, can be described by a series of equilibria:

$$S + S \xrightleftharpoons{K_2} S_2 + S \xrightleftharpoons{K_3} S_3 \cdots \xrightleftharpoons{K_n} S_n + S \longleftrightarrow \cdots \qquad (2.3.2)$$

with equilibrium constants K_n for $n = 2 - \infty$, and where the various thermodynamic parameters ($\Delta G°$, $\Delta H°$, $\Delta S°$) for the aggregation process can be expressed in terms of K_n. However, each K_n cannot be measured individually, so different approaches have been proposed to model the energetics of the process of self-association. Although not totally accurate, two simple models are generally encountered: the closed-association and the phase separation models. In the closed-association model, with the size range of spherical micelles around the CMC being very limited, it is assumed that only one of K_n value is dominant, and micelles and monomeric species are considered to be in chemical equilibrium.

$$nS \longleftrightarrow S_n \qquad (2.3.3)$$

where n is the number of molecules of surfactant, S, associating to form the micelle (i.e., the aggregation number). In the phase separation model, the micelles are considered to form a new phase within the system at and above the critical micelle concentration, and

$$nS \longleftrightarrow mS + S_n \qquad (2.3.4)$$

where m is the number of free surfactant molecules in the solution and S_n the new phase. In both cases, equilibrium between monomeric surfactant and micelles is assumed with a corresponding equilibrium constant, K_m, given by

$$K_m = \frac{[\text{micelles}]}{[\text{monomers}]^n} = \frac{[S_n]}{[S]^n} \qquad (2.3.5)$$

where brackets indicate molar concentrations and n is the number of monomers in the micelle, the aggregation number. Although micellisation is itself a source of non-ideality [31,32], it is assumed in Eq. 2.3.5 that activities may be replaced by concentrations.

From Eq. 2.3.5, the standard free energy of micellisation per mole of micelles is given by

$$\Delta G^\circ_m = -RT\ln K_m = -RT\ln[S_n] + nRT\ln[S] \qquad (2.3.6)$$

while the standard free energy change per mole of surfactant is

$$\frac{\Delta G^\circ_m}{n} = -\frac{RT}{n}\ln[S_n] + RT\ln[S] \qquad (2.3.7)$$

At (or near) the CMC, $[S] \approx [S_n]$, so that the first term on the right side of Eq. 2.3.7 can be neglected, and an approximate expression for the free energy of micellisation per mole of a neutral surfactant is

$$\Delta G^\circ_{M,m} \approx RT\ln(\text{CMC}) \qquad (2.3.8)$$

In the case of ionic surfactants, the presence of the counterion and its degree of association with the monomer and micelle must be considered. The mass-action equation becomes

$$nS^x + (n - p)C^y \leftrightarrow S_n^\alpha \quad (2.3.9)$$

where C is the concentration of free counterions. The degree of dissociation of the surfactant molecules in the micelle, α, the micellar charge, is given by $\alpha = p/n$.

The ionic equivalent to Eq. 2.3.5 is then

$$K_m = \frac{[S_n]}{[S^x]^n \times [C^y]^{(n-p)}} \quad (2.3.10)$$

where p is the concentration of free counterions associated with, but not bound to the micelle. The standard free energy of micelle formation becomes

$$\Delta G_m^\circ = -RT\{\ln[S_n] - n\ln[S^x] - (n-p)\ln[C^y]\} \quad (2.3.11)$$

At the CMC $[S^{-(+)}] = [C^{+(-)}] = $ CMC for a fully ionised surfactant, and the standard free energy change per mole of surfactant can be obtained from the approximation:

$$\Delta G_{M,m}^\circ \approx RT\left(2 - \frac{p}{n}\right)\ln(\text{CMC}) \quad (2.3.12)$$

When the ionic micelle is in a solution of high electrolyte content, the situation described by Eq. 2.3.12 reverts to the simple non-ionic case given by Eq. 2.3.8.

From the Gibbs function and second law of thermodynamics, ΔS° for non-ionic surfactants is given as

$$\Delta S^\circ = -\frac{d(\Delta G^\circ)}{dT} = -RT\frac{d\ln(\text{CMC})}{dT} - R\ln(\text{CMC}) \quad (2.3.13)$$

From the Gibbs function and Eq. 2.3.8 and 2.3.13, the enthalpy of micellisation for non-ionic surfactants, ΔH°, is given by

$$\Delta H^\circ = \Delta G^\circ + T\Delta S^\circ = -RT^2\frac{d\ln(\text{CMC})}{dT} \quad (2.3.14)$$

and similarly for ionics,

$$\Delta H^\circ = -RT^2\left(2 - \frac{p}{n}\right)\frac{d\ln(\text{CMC})}{dT} \quad (2.3.15)$$

Both the phase separation and closed association models have disadvantages. One difficulty is activity coefficients: assuming ideality can be erroneous considering the large effective micelle size and charge in comparison to dilute solutions of surfactant monomers. However, the model described above is useful enough to be applied to the systems presented in this study. Another disadvantage is the assumption of micellar monodispersity. To counteract this problem, the multiple equilibrium model was proposed, which is an extension of the closed association model. It allows a distribution function of aggregation numbers in micelles to be calculated. A full account of this model and its derivation can be found in references [33-35].

2.3.2 Factors affecting the CMC

Many factors are known to affect strongly the CMC. Of major effect is the structure of the surfactant, as will be described below. Also important, but to a lesser extent, are parameters such as counterion nature, presence of additives and change in temperature.

1. The hydrophobic group: the 'tail'

The length of the hydrocarbon chain is a major factor determining the CMC. For a homologous series of linear single-chain surfactants the CMC decreases logarithmically with carbon number. The relationship usually fits the Klevens equation [36]:

$$\lg(CMC) = A - Bn_c \qquad (2.3.16)$$

where A and B are constants for a particular homologous series and temperature, and n_c is the number of carbon atoms in the chain, C_nH_{2n+1}. The constant A varies with the nature and number of hydrophilic groups, while B is constant and approximately equal to $\lg 2$ ($B \approx 0.29$-0.30) for all paraffin chain salts having a single ionic head group (i.e., reducing the CMC to approximately one-half per each additional $-CH_2-$ group).

Interestingly, for straight-chain dialkyl sulfosuccinates Eq. 2.3.16 is still valid [37] and $B \approx 0.62$, which essentially doubles the value for the single chain compounds. Alkyl chain branching and double bonds, aromatic groups or some other polar character in the hydrophobic part produce noticeable changes in CMC. In hydrocarbon surfactants, chain branching gives a higher CMC than a comparable straight chain surfactant [19], and introduction of a benzene ring in the chain is equivalent to about 3.5 carbon atoms.

2. The hydrophilic group

For surfactants with the same hydrocarbon chain, varying the hydrophile nature (i.e., from ionic to non-ionic) has an important effect on the CMC values. For instance, for a C12 hydrocarbon the CMC with an ionic head group lies in the range of 1×10^{-3} mol·dm^{-3}, while a C12 non-ionic material exhibits a CMC in the range of 1×10^{-4} mol·dm^{-3}. The exact nature of the ionic group, however, has no dramatic effect, since a major driving force for micelle formation is the entropy factor discussed above.

3. Counterion effects

In ionic surfactants micelle formation is related to the interactions of solvent with the ionic head group. Since electrostatic repulsions between ionic groups are greatest for complete ionisation, an increase in the degree of ion binding will decrease the CMC. For a given hydrophobic tail and anionic head group, the CMC decreases as $Li^+ > Na^+ > K^+ > Cs^+ > N(CH_3)_4^+ > N(CH_2CH_3)_4^+ > Ca^{2+} \approx Mg^{2+}$. For cationic series such as the dodecyltrimethylammonium halides, the CMC decreases in the order $F^- > Cl^- > Br^- > I^-$. In addition, varying counterion valency produces a significant effect. Changing from monovalent to di-or trivalent counterions produces a sharp decrease in the CMC.

4. Effect of added salt

The presence of an indifferent electrolyte causes a decrease in the CMC of most surfactants. The greatest effect is found for ionic materials. The principal effect of the salt is to partially screen the electrostatic repulsion between the head groups and so lower the CMC. For ionics, the effect of adding electrolyte can be empirically quantified viz.

$$\lg(CMC) = -a \lg C_i + b \qquad (2.3.17)$$

Non-ionic and zwitterionic surfactants exhibit a much smaller effect and Eq. 2.3.17 does not apply.

5. Effect of temperature

The influence of the temperature on micellisation is usually weak, reflecting subtle changes in bonding, heat capacity and volume that accompany the transition. This is, however, quite a complex effect. It was shown, for example, that the CMC of most ionic surfactants passes through a minimum as the temperature is varied from 0 to 70℃ [38]. As already mentioned (Section 2.2), the major effects of temperature are the Krafft and cloud points.

2.3.3 Structure of micelles and molecular packing

Early studies [39,40] showed that, with ionic single alkyl chain compounds spherical micelles form. In particular, in 1936 Hartley [41] described such micelles as spherical aggregates whose alkyl groups form a hydrocarbon liquid-like core, and whose polar groups form a charged surface. Later, with the development of zwitterionic and non-ionic surfactants, micelles of very different shapes were encountered. The different geometries were found to depend mainly on the structure of the surfactant, as well as environmental conditions (e.g., concentration, temperature, pH, electrolyte content).

In the micellisation process, molecular geometry plays an important role and it is essential to understand how surfactants can pack. The main structures encountered are spherical micelles, vesicles, bilayers, or inverted micelles. As described previously, two opposing forces control the self-association process: hydrocarbon-water interactions that favour aggregation (i.e., pulling surfactant molecules out of the aqueous environment), and head group interactions that work in the opposite sense. These two contributions can be considered as an attractive interfacial tension term due to hydrocarbon tails and a repulsion term depending on the nature of the hydrophilic group. More recently, this basic idea was reviewed and quantified by Mitchell and Ninham [42] and Israelachvili [43], resulting in the concept that aggregation of surfactants is controlled by a balanced molecular geometry. In brief, the geometric treatment separates the overall free energy of association to three critical geometric terms (Figure 2.5):

- the minimum interfacial area occupied by the head group, a_o;
- the volume of the hydrophobic tail(s), V;
- the maximum extended chain length of the tail in the micelle core, l_c.

Formation of a spherical micelle requires l_c to be equal to or less than the micelle core radius, R_{mic}. Then for such a shape, an aggregation number, N, can be expressed either as the ratio of micellar core volume,

V_{mic}, and that for the tail, v:

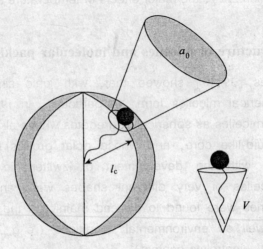

Figure 2.5 The critical packing parameter P_c (or surfactant number) relates the head group area, the extended length and the volume of the hydrophobic part of a surfactant molecule into a dimensionless number $P_c = v/a_o l_c$.

$$N = V_{mic}/v = [(4/3)\pi R_{mic}^3]/v \qquad (2.3.18)$$

or as the ratio between the micellar area, A_{mic}, and the cross-sectional area, a_o:

$$N = A_{mic}/a_o = [4\pi R_{mic}^2]/a_o \qquad (2.3.19)$$

Equating Eq. 2.3.18 and 2.3.19:

$$v/(a_o R_{mic}) = 1/3 \qquad (2.3.20)$$

Since l_c cannot exceed R_{mic} for a spherical micelle

$$v/(a_o l_o) \leq 1/3 \qquad (2.3.21)$$

More generally, this defines a critical packing parameter, P_c, as the ratio of volume to surface area:

$$P_c = v/(a_o l_c) \qquad (2.3.22)$$

The parameter v varies with the number of hydrophobic groups, chain unsaturation, chain branching and chain penetration by other compatible hydrophobic groups, while a_o is mainly governed by electrostatic interactions and head group hydration. P_c is a useful quantity since it allows the prediction of aggregate shape and size. The predicted aggregation characteristics of surfactants cover a wide range of geometric possibilities, and the main types are presented in Table 2.1 and Figures 2.6 and 2.7.

Table 2.1 **Expected aggregate characteristics in relation to surfactant critical packing parameter**, $P_c = v/a_o l_c$

P_c	General Surfactant Type	Expected Aggregate Structure
< 0.33	Single-chain surfactants with large head groups	Spherical or ellipsoidal micelles
0.33—0.5	Single-chain surfactants with small head groups, or ionics in the presence of large amounts of electrolyte	Large cylindrical or rod-shaped micelles Vesicles and flexible bilayers structures
0.5—1.0	Double-chain surfactants with large head groups and flexible chains	Planar extended bilayers
1.0	Double-chain surfactants with small head groups or rigid, immobile chains	Reversed or inverted micelles
> 1.0	Double-chain surfactants with small head groups, very large and bulky hydrophobic groups	

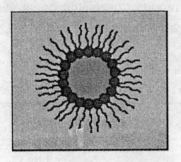 

Negative or reversed curvatuer
P>1
Water-in-oil
Oil-soluble micelles

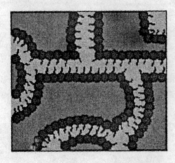 

Zero or planar curvature
P~1
Bicontinuous

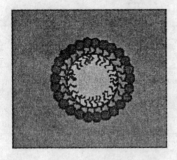 

Positive or mormal curature
P<1
Water-soluble micelles
Oil-in-water microemulsions

Figure 2.6 Changes in the critical packing parameters (P_c) of surfactant molecules give rise to different aggregation structures.

2.4 LIQUID CRYSTALLINE MESOPHASES

Micellar solutions, although the subject of extensive studies and theoretical considerations, are only one of several possible aggregation states. A complete understanding of the aqueous behaviour of surfactants requires knowledge of the entire spectrum of self-assembly. The existence of liquid crystalline phases constitutes an equally important aspect and a detailed description can be found in the literature [e.g. ,44, 45]. The common features of liquid crystalline phases are summarised below.

2.4.1 Definition

When the volume fraction of surfactant in a micellar solution is increased, typically above a threshold of about 40%, a series of regular geometries is commonly encountered. Interactions between micellar surfaces are repulsive (from electrostatic or hydration forces), so that as the number of aggregates increases and micelles get closer to one another, the only way to maximise separation is to change shape and size. This explains the sequence of surfactant phases observed in the concentrated regime. Such phases are known as mesophases or lyotropic (solvent-induced) liquid crystals.

As the term suggests, liquid crystals are characterised by having physical properties intermediate between crystalline and fluid structures: the degree of molecular ordering is between that of a liquid and a crystal and in terms of rheology the systems are neither simple viscous liquids nor crystalline elastic solids. Certain of these phases have at least one direction that is highly ordered so that liquid crystals exhibit optical birefringence.

Two general classes are encountered depending on whether one is considering surfactants or other types of material. These are *thermotropic* liquid crystals, in which the structure and properties are determined by temperature (such as employed in LCD cells). For *lyotropic* liquid crystals structure is determined by specific interactions between solute and solvent: surfactant liquid crystals are normally lyotropic.

2.4.2 Structures

The main structures associated with two-component surfactant-water systems are: hexagonal (normal or inverted), lamellar, and several cubic phases. Table 2.2 summarises the notations commonly associated with these phases and their structures are shown in Figure 2.6.

- The *hexagonal phase* is composed of a close-packed array of long cylindrical micelles, arranged in a hexagonal pattern. The micelles may be "normal" (in water, H_1) in that the hydrophilic head groups are located on the outer surface of the cylinder, or "inverted" (H_2), with the hydrophilic group located internally. Since all the space between adjacent cylinders is filled with hydrophobic groups, the cylindrical micelles are more closely packed than those found in the H_1 phase. As a result, H_2 phases occupy a much smaller region of the phase diagram and are much less common.
- The *lamellar phase* (L_α) is built up of alternating water-surfactant bilayers. The hydrophobic chains possess a significant degree of randomness and mobility, and the surfactant bilayer can range from being stiff and planar to being very flexible and undulating. The level of disorder may vary smoothly or change abruptly, depending on the specific system, so that it is possible for a surfactant to pass through several distinct lamellar phases.
- The *cubic phase* may have a wide variety of structural variations and occurs in many different parts of the phase diagram. They are optically isotropic systems and so cannot be characterised by polarised light microscopy. Two main groups of cubic phases have been identified:
  i. The micellar cubic phases (I_1 and I_2) —built up of regular packing of small micelles (or reversed micelles in the case of I_2). The micelles are short prolates arranged in a body — centred cubic close-packed array [46,47].
  ii. The bicontinuous cubic phases (V_1 and V_2)—are thought to be rather extended, porous, connected structures in three dimensions. They are considered to be formed by either connected rod — like

micelles, similar to branched micelles, or bilayer structures. Denoted V_1 and V_2, they can be normal or reverse structures and are positioned between H_1 and L_α and between L_α and H_2 respectively.

In addition to having different structures these common forms also show different viscosities, in the order:

$$\text{Cubic} > \text{Hexagonal} > \text{Lamellar}$$

Cubic phases are generally the more viscous since they have no obvious shear plane and so layers of surfactant aggregates cannot slide easily relative to each other. Hexagonal phases typically contain 30%—60% water by weight but are very viscous since cylindrical aggregates can move freely only along their length. Lamellar phases are generally less viscous than the hexagonal phases due to the ease with which each parallel layers can slide over each other during shear.

Table 2.2 **Most common lyotropic liquid crystalline and other phases found in binary surfactant-water systems.**

Phase structure	Symbol	Other names
Lamellar	L_α	Neat
Hexagonal	H_1	Middle
Reversed hexagonal	H_2	
Cubic (normal micellar)	I_1	Viscous isotropic
Cubic (reversed micellar)	I_2	
Cubic (normal bicontinuous)	V_1	Viscous isotropic
Cubic (reversed bicontinuous)	V_2	
Micellar	L_1	
Reversed micellar	L_2	

2.4.3 Phase diagrams

The sequence of mesophases can be identified simply by using a polarising microscope and the isothermal technique known as a phase cut. Briefly, starting from a small amount of surfactant, a concentration gradient

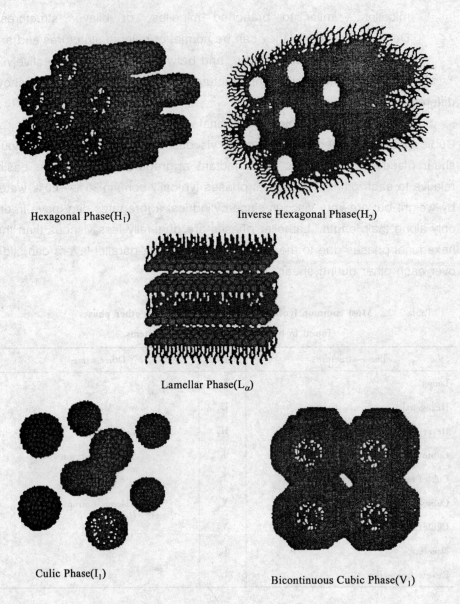

Figure 2.7 common surfactant liquid crystalline phases. See Table 2.2 for identification.

is set up spanning the entire phase diagram, from pure water to pure surfactant. Since crystal hydrates and some of the liquid crystalline phases are birefringent, viewing in the microscope between crossed polars shows up the complete sequence of mesophases.

Transformations between different mesophases are controlled by a balance between molecular packing geometry and inter-aggregate forces. As a result, the system characteristics are highly dependent on the nature and amount of solvent present. Generally, the main types of mesophases tend to occur in the same order and in roughly the same position in the phase diagram. Figure 2.8 shows a classic binary phase diagram of a non-ionic surfactant $C_{16}EO_8$-water. The sequence of phases is common to most non-ionic surfactants of the kind C_iE_j, although the positions of the phase boundaries, in terms of temperature and concentration limits, depend somewhat on the chemical identity of the surfactant.

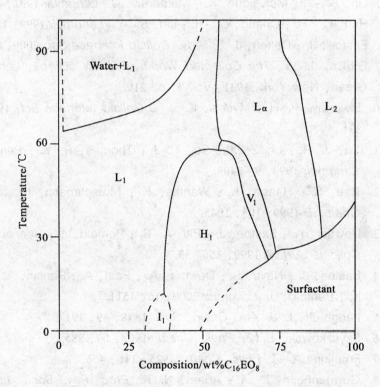

Figure 2.8 Phase diagram for the non-ionic $C_{16}EO_8$ illustrating the various liquid crystalline phases. L_1 and L_2 are isotropic solutions. See Table 2.2 for details of the other phases. (After Mitchell *et al. J. Chem. Soc. Faraday Trans. I* 1983, 79. 975) Reproduced by permission of The Royal Society of Chemistry).

REFERENCES

1. Tanford, C. '*The Hydrophobic Effect*: *formation of micelles and biological membranes*', John Wiley & Sons, 1978, USA.
2. Dukhin, S. S.; Kretzschmar, G.; Miller, R. '*Dynamics of Adsorption at Liquid Interfaces*', Elsevier, Amsterdam, 1995.
3. Rusanov, A. I.; Prokhorov, V. A. *Interfacial Tensiometry*, Elsevier, Amsterdam, 1996.
4. Chang, C.-H.; Franses, E. I. *Colloid Surf.* 1995, 100, 1.
5. Miller, R.; Joos, P.; Fainermann, V. *Adv. Colloid Interface Sci.*, 1994, 49, 249.
6. Lin, S.-Y.; McKeigue, K.; Maldarelli, C. *Langmuir* 1991, 7, 1055.
7. Hsu, C.-H.; Chang, C.-H.; Lin, S.-Y. *Langmuir* 1999, 15, 1952.
8. Eastoe, J.; Dalton, J. S. *Adv. Colloid Interface Sci.* 2000, 85, 103.
9. Gibbs, J. W. *The Collected Works of J. W. Gibbs*, Longmans, Green, New York, 1931, Vol. I, p. 219.
10. Elworthy, P. H.; Mysels, K. J. *J Colloid Interface Sci.* 1966, 21, 331
11. Lu, J. R.; Li, Z. X.; Su, T. J.; Thomas, R. K.; Penfold, J. *Langmuir* 1993, 9, 2408.
12. Bae, S.; Haage, K.; Wantke, K.; Motschmann, H. *J. Phys. Chem. B* 1999, 103, 1045.
13. Downer, A.; Eastoe, J.; Pitt, A. R.; Penfold, J.; Heenan, R. K. *Colloids Surf. A* 1999, 156, 33.
14. Eastoe, J.; Nave, S.; Downer, A.; Paul, A.; Rankin, A.; Tribe, K.; Penfold, J. *Langmuir* 2000, 16, 4511.
15. Langmuir, I. *J. Am. Chem. Soc.* 1848, 39, 1917.
16. Szyszkowski, B. *Z. Phys. Chem.* 1908, 64, 385.
17. Frumkin, A. *Z. Phys. Chem.* 1925, 116, 466.
18. Guggenheim, E. A.; Adam, N. K. *Proc. Roy. Soc. (London)*, 1933, A139, 218.
19. Rosen, M. J. '*Surfactants And Interfacial Phenomena*', John Wiley & Sons, 1989, USA.
20. Traube, I. *Justus Liebigs Ann. Chem.* 1891, 265, 27.

21. Tamaki, K. ; Yanagushi, T. ; Hori, R. *Bull. Chem. Soc. Jpn.* 1961, *34*, 237.
22. Pitt, A. R. ; Morley, S. D; Burbidge, N. J. ; Quickenden, E. L. *Coll. Surf. A* 1996, *114*, 321.
23. Hato, M. ; Tahara, M. ; Suda, Y. *J. Coll. Interface Sci.* 1979, *72*, 458.
24. Staples, E. J. ; Tiddy, G. J. T. *J. Chem. Soc., Faraday Trans. 1* 1978, *74*, 2530
25. Tiddy, G. J. T. *Phys. Rep.* 1980, *57*, 1.
26. Schott, H. *J. Pharm. Sci.* 1969, *58*, 1443.
27. Frank, H. S. ; Evans, M. W. *J. Chem. Phys.* 1945, *13*, 507.
28. Evans, D. F. ; Wightman, P. J. *J. Colloid Interface Sci.* 1982, *86*, 515.
29. Patterson, D. ; Barbe, M. *J. Phys. Chem.* 1976, 80, 2435.
30. Evans, D. F. *Langmuir* 1988, 4, 3.
31. Hunter, R. J. '*Foundations of Colloid Science Volume I*', Oxford University Press, 1987, New York.
32. Evans, D. F. ; Ninham, B. W. *J. Phys. Chem.* 1986, *90*, 226.
33. Corkhill, J. M. ; Goodman, J. F. ; Walker, T. ; Wyer, J. *Proc. Roy. Soc. (London), A* 1969, *312*, 243.
34. Mukerjee, P. *J. Phys. Chem.* 1972, 76, 565.
35. Aniansson, E. A. G. ; Wall, S. N. *J. Phys. Chem.* 1974, *78*, 1024.
36. Klevens, H. *J. Am. Oil Chem. Soc.* 1953, *30*, 7, 4.
37. Williams, E. F. ; Woodberry, N. T. ; Dixon, J. K. *J. Colloid Interface Sci.* 1957, *12*, 452.
38. Kresheck, G. C. In *Water-a comprehensive treatise*, pp. 95-167. Ed. F. Franks, Plenum Press, 1975, New York.
39. McBain, J. W. *Trans. Faraday Soc.* 1913, *9*, 99.
40. Reychler, *Kolloid-Z.* 1913, *12*, 283.
41. Hartley, G. S. '*Aqueous solutions of paraffin chain salts*', Hermann & Cie, Paris, 1936.
42. Mitchell, D. J. ; Ninham, B. W. *J. Chem. Soc. Faraday Trans. 2* *1981*, *77*, 601.

43. Israelachvili, J. N. '*Intermolecular and Surface Forces*', Academic Press, London, 1985, p. 251.
44. Laughlin, R. G. '*The Aqueous Phase Behaviour of Surfactants*', Academic Press, London, 1994.
45. Chandrasekhar, S. '*Liquid Crystals*', Cambridge University Press, 1992, New York.
46. Fontell, K.; Kox, K. K.; Hansson, E. *Mol. Cryst. Liquid Cryst. Letters* 1985, *1*, 9.
47. Fontell, K. Coll. *Polymer Sci.* 1990, 268, 264.

Appendix 1 - Tensiometric Methods

Tensiometry is a very accessible method but only provides indirect determination of the surface excess via surface tension measurements and application of the Gibbs equation (see Section 2.1.2. equations 2.1.13 and 2.1.15). Below, the main features of drop volume and du Noüy ring tensiometry techniques are described.

Most techniques for measuring equilibrium surface tension involve stretching the liquid-air interface at the moment of measurement. Equilibrium surface tension can be obtained by measuring a force, pressure or drop size. The ring and plate methods both measure a force, whereas the capillary height and maximum bubble pressure methods rely on pressure. The pendant drop, sessile drop, drop volume, drop weight and spinning drop methods all measure one or more dimensions of a drop.

A.1 DU NOüY RING TENSIOMETRY

The ring method [A1-A4] involves a platinum-iridium ring, attached to a vertical wire, being immersed horizontally into the liquid, see figure A.1 below.

The surface tension is calculated from the force required to pull the ring through the interface. Assuming the ring supports a cylinder of liquid, the surface tension is given by

$$\gamma_{eq} = \frac{F}{4\pi R} \tag{A.1}$$

2. Aggregation and adsorption at interfaces

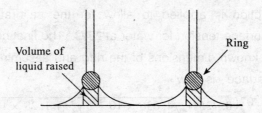

Figure A.1 Schematic of Du Noüy ring

where R is the radius of the ring. At equilibrium the maximum force is given by

$$F = (\rho_1 - \rho_2)gV \qquad (A.2)$$

where ρ_1 and ρ_2 are densities of the liquid phase and the liquid or gas phase above it, g is acceleration due to gravity ($9.81 \text{ m} \cdot \text{s}^{-2}$), and V is the volume of liquid raised by the ring. For a dilute aqueous solution-air interface, ρ_1 is assumed to be the density of water, and ρ_2, the density of air, so by measuring the weight of the liquid raised above the surface, the surface tension can be calculated.

However, the main disadvantage of the ring method is that a correction factor is required. This is because the liquid column lifted by the ring is not quite a cylinder, and that the balance measures the weight of the water lifted. This correction factor has been determined by Harkins and Jordan [A.1] and is incorporated as follows

$$\gamma_{eq} = \gamma_{eq}^* \cdot f = \frac{F}{4\pi R} \cdot f \qquad (A.3)$$

where f is the dimensionless Harkins and Jordan Factor and γ_{eq} the measured value in $\text{mN} \cdot \text{m}^{-1}$

The correction factor can be determined by the equation published by Zuidema and Waters, based on an interpolation of the Harkins and Jordan correction factor tables (see A.4).

$$f = 0.725 + \sqrt{\frac{0.014,52 \cdot \gamma_{eq}^*}{\frac{1}{4}U^2(\rho_1 - \rho_2)} + 0.045,34 - \frac{1.679}{R/r}} \qquad (A.4)$$

where R is the mean ring radius (typically 10 mm), r is the radius of the cross-section of the wire (typically 0.2 mm), U is the wetting length (typically 120 mm).

A final correction is applied to allow for the calibration, done with reference to the surface tension for water at 20°C. The final correction factor, after inserting the known dimensions of the ring and assuming $(\rho_1 - \rho_2) = 1$ for a water-air interface, is now

$$fk = 1.07 \left(0.725 + \sqrt{4.036 \times 10^{-4} \cdot \gamma_{eq}^* + 1.28 \times 10^{-2}} \right) \quad (A.5)$$

A.2 DROP VOLUME TENSIOMETRY—DVT

The principle behind DVT is the determination of the maximum size of a drop formed at the end of a well-defined capillary. A modern commercial rig (e.g. Lauda TVT1 drop volume tensiometer) is fully automated and sophisticated dosing regimes can be selected so that dynamic surface tension may be followed. A full description of this method is given elsewhere [A2, A3]. Briefly, as shown in figure A.2, the stepper motor lowers a

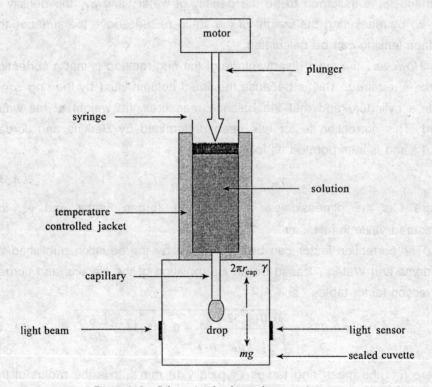

Figure A.2 Schematic of a drop volume tensiometer

barrier onto a syringe plunger and causes a drop to form at the capillary tip. As the stepper motor continues the drop will grow until the weight of the drop acting downward (mg) exceeds the tension force acting upward ($2\pi r_{cap}\gamma$). The drop will then detach from the capillary and a light sensor detects this movement. Hence the maximum volume of the drop, V, is related to the surface tension, γ, via equation A.6 [A.4]:

$$\gamma = \frac{V\Delta\rho g}{2\pi r_{cap}} f \qquad (A.6)$$

where $\Delta\rho$ is the density difference between the two phases, g is the acceleration due to gravity, and r_{cap} is the capillary radius; f is a correction factor accounting for the point of drop detachment being not at the capillary tip but at its own neck [A5].

A.3 CALCULATION OF ACTIVITY COEFFICIENTS

When studying the surface tension-concentration behaviour of ionic surfactants, activity rather than concentration should be used. Whilst in very dilute solution, i.e., below 1×10^{-3} mol dm^{-3}, activity coefficients can safely be regarded as unity, at higher concentrations, i.e., above 1×10^{-3} moldm^{-3}, this assumption is no longer valid. Coulombic interactions between ions increase result in departure from ideal behaviour and require the use of Debye-Hückel theory to consider the effect of ionic strength. This is explained in detail in standard texts [A6, A7] and only relevant equations are given here. At very low electrolyte concentrations, the mean activity coefficient $\gamma_\pm$ can be calculated from the Debye-Hückel limiting law:

$$\lg\gamma_\pm = -A |z_+ z_-| I^{1/2} \qquad (A.7)$$

where z is the charge on the ion, I is the ionic strength and A is a constant. The form of I and the constant A are given below

$$I = \frac{1}{2}\sum_i m_i Z_i^2 \qquad (A.8)$$

$$A = \frac{F^3}{4\pi N_a \ln 10}\left(\frac{\rho}{2(\varepsilon_o \varepsilon_r RT)^3}\right)^{1/2} \qquad (A.9)$$

where m is the molality, z is the charge valency, and ρ is the solvent density. F, N_a, R, ε_o and ε_r are all standard physical constants.

For 1:1 electrolytes equation (A.7) is valid for concentrations below approximately 0.01mol dm^{-3}. For other valence types, or higher concentrations, the Debye-Hückel extended law must be used:

$$\lg\gamma_{\pm} = -\frac{A|z_+ z_-|I^{1/2}}{1 + BaI^{1/2}} \qquad (A.10)$$

where a is the mean effective ionic diameter which typically ranges from 3-9Å [A8] and B is a constant given by

$$B = \left(\frac{2F^2\rho}{\varepsilon_o\varepsilon_r RT}\right)^{1/2} \qquad (A.11)$$

Equation (A.10) extends the validity of Debye-Hückel theory for 1:1 electrolytes up to concentrations of $0.1 \text{ mol} \cdot \text{dm}^{-3}$ [A7]. For aqueous solutions at 298 K, $A = 0.509 \text{ mol}^{-1/2} \cdot \text{kg}^{1/2}$ and $B = 3.282 \times 10^9 \text{ m}^{-1} \cdot \text{mol}^{-1/2} \cdot \text{kg}^{1/2}$.

A.1 REFERENCES

A1. Harkins, W. D.; Jordan, H. F., *J. Am. Chem. Soc.* 1930, 52, 1751.

A2. Miller, R.; Joos, P.; Fainerman, V. B., *Adv. Coll. Int. Sci.* 1994, 49, 249.

A3. Dukhin, S. S.; Kretzschmar, G.; Miller, R., *Dynamics of Adsorption at Liquid Interfaces* 1995 (Amsterdam: Elsevier).

A4. Rusanov, A. I.; Prokhorov, V. A., *'Interfacial Tensiometry'*, Eds. Möbius, D.; Miller, R. 1996 (Amsterdam: Elsevier).

A5. Miller, R.; Schano, K-H.; Hofmann, A., *Colloids Surf. A* 1994, 92, 33.

A6. Atkins, P. W., '*Physical Chemistry*' 6th edition, 1998, (Oxford University Press: Oxford).

A7. Robbins, J., '*Ions in Solution*', 1972 (Oxford University Press, Oxford).

A8. Levine, I. N. '*Physical Chemistry*', 4th edition, 1995 (McGraw-Hill Book Co.; Singapore).

3. Microemulsions

This chapter is devoted to another important property of surfactants, that of stabilization of water-oil films and formation of microemulsions. These are a special kind of colloidal dispersion that have attracted a great deal of attention because of their ability to solubilise otherwise insoluble materials. Industrial applications of microemulsions have escalated in the last 40 years following an increased understanding of formation, stability and the role of surfactant molecular architecture. This chapter reviews main theoretical features relevant to the present work and some common techniques used to characterize microemulsion phases.

3.1 MICROEMULSIONS: DEFINITION AND HISTORY

One of the best definitions of microemulsions is from Danielsson and Lindman [1] "*a microemulsion is a system of water, oil and an amphiphile which is a single optically isotropic and thermodynamically stable liquid solution*". In some respects, microemulsions can be considered as small-scale versions of emulsions, i.e., droplet type dispersions either of oil-in-water (O/W) or of water-in-oil (W/O), with a size range in the order of 5-50 nm in drop radius. Such a description, however, lacks precision since there are significant differences between microcmulsions and ordinary emulsions (or macroemulsions). In particular, in emulsions the average drop size grows continuously with time so that phase separation ultimately occurs under gravitational force, i.e., they are thermodynamically unstable and their formation requires input of work. The drops of the dispersed phase are generally large (> 0.1 μm) so that they often take on a milky rather than a translucent appearance. For microemulsions, once the conditions are right, spontaneous formation occurs. As for simple aqueous systems, microemulsion formation is dependent on surfactant type and structure. If the surfactant is ionic and contains a single hydrocarbon chain (e.g., sodium dodecylsulphate, SDS) microemulsions are only formed if a co-surfactant (e.g., a medium size aliphatic alcohol) and/or electrolyte (e.g., 0.2 M NaCl) are also present. With double chain ionics (e.g., Aerosol-OT) and some non-ionic surfactants a co-surfactant is not necessary. This results from one of the most fundamental properties of microemulsions, that is, an ultra-low interfacial tension between the oil and water phases, $\gamma_{o/w}$: the main role of the surfactant is to reduce $\gamma_{o/w}$ sufficiently —i.e., lowering the energy required to increase the surface area— so that spontaneous dispersion of water or oil droplets occurs and the system is thermodynamically stable. As described in Section 3.2.1 ultra-low tensions are crucial for the formation of microemulsions and depend on system composition.

Microemulsions were not really recognized until the work of Hoar and Schulman in 1943, who reported a spontaneous emulsion of water and oil on addition of a strong surface-active agent [2]. The term "microemulsion" was first used even later by Schulman *et al.* [3] in 1959 to describe a multiphase system consisting of water, oil, surfactant and alcohol, which forms a transparent solution. There has been much debate about the word "microemulsion" to describe such systems [4]. Although not systematically used today, some prefer the names "micellar emulsion" [5] or "swollen micelles" [6]. Microemulsions were probably discovered well before the studies of Schulmann: Australian housewives have used since the beginning of last century water/eucalyptus oil/soap flake/white spirit mixtures to wash wool, and the first commercial microemulsions were probably the liquid waxes discovered by Rodawald in 1928. Interest in microemulsions really stepped up in the late 1970's and early 1980's when it was recognized that such systems could improve oil recovery and when oil prices reached levels where tertiary recovery methods became profit earning [7]. Nowadays this is no longer the case, but other microemulsion applications were discovered, e.g., catalysis, preparation of submicron particles, solar energy conversion, liquid-liquid extraction (mineral, proteins, etc.). Together with classical applications in detergency and lubrication, the field remains sufficiently important to continue to attract a number of scientists. From the fundamental research point of view, a great deal of progress has been made in the last 20 years in understanding microemulsion properties. In particular, interfacial film stability and microemulsion structures can now be characterized in detail owing to the development of new and powerful techniques such as small-angle neutron scattering (SANS), as described in Chapter 4). The following sections deal with fundamental microemulsion properties, i. e., formation and stability, surfactant films, classification and phase behaviour.

3.2 THEORY OF FORMATION AND STABILITY

3.2.1 Interfacial tension in microemulsions

A simple picture for describing microemulsion formation is to consider a subdivision of the dispersed phase into very small droplets. Then the configurational entropy change, ΔS_{conf}, can be approximately expressed as [8]

$$\Delta S_{conf} = -nk_B[\ln\phi + \{(1-\phi)/\phi\}\ln(1-\phi)] \quad (3.2.1)$$

where n is the number of droplets of dispersed phase, k_B is the Boltzmann constant and ϕ is the dispersed phase volume fraction. The associated free energy change can be expressed as a sum of the free energy for creating new area of interface, $\Delta A\gamma_{12}$, and configurational entropy in the form [9]:

$$\Delta G_{form} = \Delta A\gamma_{12} - T\Delta S_{conf} \quad (3.2.2)$$

where ΔA is the change in interfacial area A (equal to $4\pi r^2$ per droplet of radius r) and γ_{12} is the interfacial tension between phases 1 and 2 (e. g., oil and water) at temperature T (Kelvin). Substituting Eq. 3.2.1 into 3.2.2 gives an expression for obtaining the maximum interfacial tension between phases 1 and 2. On dispersion, the droplet number increases and ΔS_{conf} is positive. If the surfactant can reduce the interfacial tension to a sufficiently low value, the energy term in Eq. 3.2.2 ($\Delta A\gamma_{12}$) will be relatively small and positive, thus allowing a negative (and hence favourable) free energy change, that is, spontaneous microemulsification.

In surfactant-free oil-water systems, $\gamma_{o/w}$ is of the order of 50 mN · m^{-1}, and during microemulsion formation the increase in interfacial area, ΔA, is very large, typically a factor of 10^4 to 10^5. Therefore in the absence of surfactant, the second term in Eq. 3.2.2 is of the order of 1,000 $k_B T$, and in order to fulfill the condition $\Delta A \gamma_{12} \leq T \Delta S_{conf}$, the interfacial tension should be very low (approximately 0.01 mN $\cdot$ m^{-1}). Some surfactants (double chain ionics [10, 11] and some non-ionics [12]) can produce extremely low interfacial tensions— typically 10^{-2} to 10^{-4} mN $\cdot$ m^{-1}—but in most cases, such low values cannot be achieved by a single surfactant since the CMC is reached before a low value of $\gamma_{o/w}$ is attained. An effective way to further decrease $\gamma_{o/w}$ is to include a second surface-active species (either a surfactant or medium-chain alcohol), that is a co-surfactant. This can be understood in terms of the Gibbs equation extended to multicomponent systems [13]. It relates the interfacial tension to the surfactant film composition and the chemical potential, μ, of each component in the system, i.e.,

$$d\gamma_{o/w} = -\sum_i (\Gamma_i d\mu_i) \approx -\sum_i (\Gamma_i RT d\ln C_i) \qquad (3.2.3)$$

where C_i is the molar concentration of component i in the mixture, and Γ_i the surface excess (mol $\cdot$ m^{-2}). Assuming that surfactants and co-surfactants, with concentration C_s and C_{co} respectively, are the only adsorbed components (i.e., $\Gamma_{water} = \Gamma_{oil} = 0$), Eq. 3.2.3 becomes

$$d\gamma_{o/w} = -\Gamma_s RT d\ln C_s - \Gamma_{co} RT d\ln C_{co} \qquad (3.2.4)$$

Integration of Eq. 3.2.4 gives

$$\gamma_{o/w} = \gamma^\circ_{o/w} - \int_0^{C_s} \Gamma_s RT d\ln C_s - \int_0^{C_{co}} \Gamma_{co} RT d\ln C_{co} \qquad (3.2.5)$$

Eq. 3.2.5 shows that $\gamma^\circ_{o/w}$ is lowered by two terms, both from the surfactant and co-surfactant (of surface excesses Γ_s and Γ_{co} respectively) so their effects are additive. It should be mentioned, however, that the two molecules should be adsorbed simultaneously and should not interact with each other (otherwise they lower their respective activities), i.e., are of completely different chemical nature, so that mixed micellisation does not occur. With certain surfactant-co-surfactant systems the interfacial tension can become so low that further increase in concentration is not possible without making $\gamma_{o/w}$ negative. Overbeek [14] studied variation of $\gamma_{o/w}$ with concentration for the system brine-cyclohexane-*n*-pentanol-SDS (Figure

3.1): a transient negative $\gamma_{o/w}$ value is reached implying spontaneous expansion of the interface by taking up the excess of surfactant and co-surfactant. This is illustrated in Figure 3.1 with the plot for 20% pentanol.

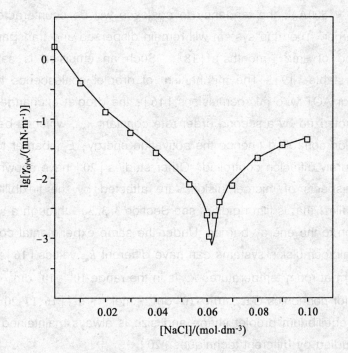

Figure 3.1 Oil-water interfacial tension between n-heptane and aqueous NaCl solutions as a function of salt concentration in the presence of AOT surfactant. The values were determined by spinning drop tensiometry. The AOT surfactant concentration is 0.050 mol · dm^{-3}, temperature 25℃.

Then $\gamma_{o/w}$ increases back to a small positive equilibrium value, generating a microemulsion spontaneously. Such thermodynamic treatment was originated by Ruckenstein and Chi [15] and Overbeek [8]. Equations 3.2.1 and 3.2.2 are very approximate; in particular, additional terms should be included to account for any specific interactions between droplets [15].

3.2.2 Kinetic stability

Internal contents of the microemulsion droplets are known to exchange, typically on the millisecond time scale [16,17]. They diffuse and undergo collisions. If collisions are sufficiently violent, then the surfactant film may rupture thereby facilitating droplets exchange, that is the droplets are kinetically unstable. However, if one disperses emulsions sufficiently small droplets (< 500Å), the tendency to coalesce will be counteracted by an energy barrier. Then the system will remain dispersed and transparent for a long period of time (months) [18]. Such an emulsion is said to be kinetically stable [19]. The mechanism of droplet coalescence has been reported for AOT w/o microemulsions [16]; the droplet exchange process was characterized by a second order rate constant k_{ex}, which is believed to be activation controlled (hence the activation energy, E_a, barrier to fusion) and not purely diffusion controlled. Other studies [20] have shown that the dynamic aspects of microemulsions are affected by the flexibility of the interfacial film, that is film rigidity (see Section 3.3.3), through a significant contribution to the energy barrier. Under the same experimental conditions, different microemulsion systems can have different k_{ex} values [16]: for AOT w/o system at room temperature, k_{ex} is in the range $10^6 \sim 10^9$ dm$^3 \cdot$ mol$^{-1} \cdot$ s^{-1}, and for non-ionics C_iE_j, $10^8 \sim 10^9$ dm$^3 \cdot$ mol$^{-1} \cdot$ s^{-1} [16,17,20]. In any case, an equilibrium droplet shape and size is always maintained and this can be studied by different techniques [20].

3.3 PHYSICOCHEMICAL PROPERTIES

This section gives an overview of the main parameters characterizing microemulsions. References will be made to related behaviour for planar interfaces presented in Chapter 2.

3.3.1 Predicting microemulsion type

A well-known classification of microemulsions is that of Winsor [21] who identified four general types of phase equilibria:
- Type I: the surfactant is preferentially soluble in water and oil-in-water (O/W) microemulsions form (Winsor I). The surfactant-rich water phase coexists with the oil phase where surfactant is only

present as monomers at small concentration.
- Type II : the surfactant is mainly in the oil phase and water-in-oil (W/O) microemulsions form. The surfactant-rich oil phase coexists with the surfactant-poor aqueous phase (Winsor II).
- Type III : a three-phase system where a surfactant-rich middle-phase coexists with both excess water and oil surfactant-poor phases (Winsor III or middle-phase microemulsion).
- Type IV : a single-phase (isotropic) micellar solution, that forms upon addition of a sufficient quantity of amphiphile (surfactant plus alcohol).

Depending on surfactant type and sample environment, types I, II, III or IV form preferentially, the dominant type being related to the molecular arrangement at the interface (see below). As illustrated in Figure 3.2, phase transitions are brought about by increasing either electrolyte concentration (in the case of ionic surfactants) or temperature (for nonionics). Table 3.1 summarizes the qualitative changes in phase behaviour of anionic surfactants when formulation variables are modified [22].

Various investigators have focused on interactions in an adsorbed interfacial film to explain the direction and extent of interfacial curvature. The first concept was that of Bancroft [23] and Clowes [24] who considered the adsorbed film in emulsion systems to be duplex in nature, with an inner and an outer interfacial tension acting independently [25]. The interface would then curve such that the inner surface was one of higher tension. Bancroft's rule was stated as "*that phase will be external in which the emulsifier is most soluble*"; i.e., oil-soluble emulsifiers will form w/o emulsions and water — soluble emulsifiers o/w emulsions. This qualitative concept was largely extended and several parameters have been proposed to quantify the nature of the surfactant film. They are briefly presented in this section. Further details concerning the three microemulsion types and their location in the phase diagram will be given in Section 3.3.3.

1. The R-ratio

The *R*-ratio was first proposed by Winsor [21] to account for the influence of amphiphiles and solvents on interfacial curvature. The primary concept is to relate the energies of interaction between the amphiphile layer and the oil and water regions. Therefore, this *R*-ratio compares the tendency for an amphiphile to disperse into oil, to its tendency to dissolve in water. If one phase is favoured, the interfacial region tends to take on a definite curvature. A brief description of the concept is given below, and a full account can be found elsewhere [26].

In micellar or microemulsion solutions, three distinct (single or multicomponent) regions can be recognized: an aqueous region, W, an oil or organic region, O, and an amphiphilic region, C. As shown in Figure 3.3, it is useful to consider the interfacial zone as having a definite composition, separating essentially bulk-phase water from bulk-phase oil. In this simple picture, the interfacial zone has a finite thickness, and will contain, in addition to surfactant molecules, some oil and water.

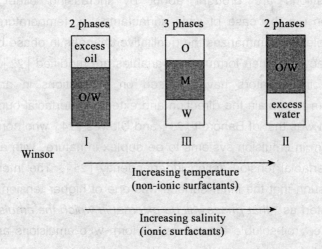

Figure 3.2 Winsor classification and phase sequence of microemulsions encountered as temperature or salinity is scanned for non-ionic and ionic surfactant respectively. Most of the surfactant resides in the shaded area. In the three-phase system the middle-phase microemulsion (M) is in equilibrium with both excess oil (O) and water (W).

Table 3.1 **Qualitative effect of several variables on the observed phase behaviour of anionic surfactants. After Bellocq *et al.* [22]**

Scanned variables (increase)	Ternary diagram transition
Salinity	I < III < II
Oil: Alkane carbon number	II < III < I

128

3. Microemulsions

Table 3.1 comtinue

Scanned variables (increase)	Ternary diagram transition
Alcohol: low M. W. [a]	I < III < II
high M. W. [b]	I < III < II
Surfactant: lipophilic chain length	I < III < II
Temperature	II < III < I

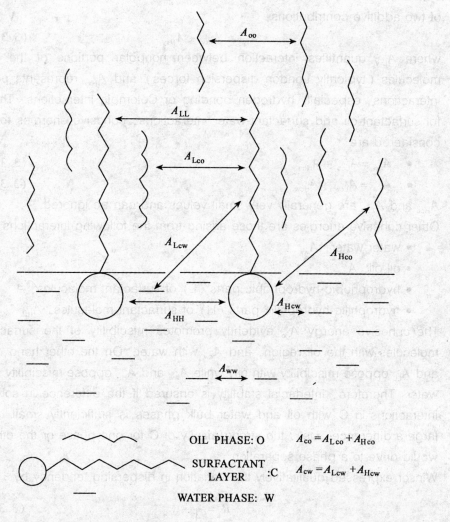

Figure 3.3 Interaction energies in the interfacial region of an oil-surfactant-water system.

$A_{co} = A_{Lco} + A_{Hco}$

$A_{cw} = A_{Lcw} + A_{Hcw}$

OIL PHASE: O
SURFACTANT LAYER :C
WATER PHASE: W

Cohesive interaction energies therefore exist within the C layer, and these determine interfacial film stability. They are depicted schematically in Figure 3.3: the cohesive energy between molecules x and y is defined as A_{xy}, and is positive whenever interaction between molecules is attractive. A_{xy} is depicted as the cohesive energy per unit area between surfactant, oil and water molecules residing in the anisotropic interfacial C layer. For surfactant-oil and surfactant-water interactions A_{xy} can be considered to be composed of two additive contributions:

$$A_{xy} = A_{Lxy} + A_{Hxy} \quad (3.3.1)$$

where A_{Lxy} quantifies interaction between nonpolar portions of the two molecules (typically London dispersion forces) and A_{Hxy} represents polar interactions, especially hydrogen bonding or Colombic interactions. Thus, for surfactant-oil and surfactant-water interactions, cohesive energies to be considered are

- $A_{co} = A_{Lco} + A_{Hco}$ (3.3.2)
- $A_{cw} = A_{Lcw} + A_{Hcw}$ (3.3.3)

A_{Hco} and A_{Lcw} are generally very small values and can be ignored.
Other cohesive energies are those arising from the following interactions:

- water-water, A_{ww}
- oil-oil, A_{oo}
- hydrophobic-hydrophobic parts (L) of surfactant molecules, A_{LL}
- hydrophilic-hydrophilic parts (H) of surfactant molecules, A_{HH}

The cohesive energy A_{co} evidently promotes miscibility of the surfactant molecules with the oil region, and A_{cw} with water. On the other hand, A_{oo} and A_{LL} oppose miscibility with oil, while A_{ww} and A_{HH} oppose miscibility with water. Therefore, interfacial stability is ensured if the difference in solvent interactions in C with oil and water bulk phases is sufficiently small. Too large a difference, i.e., too strong affinity of C for one phase or the other, would drive to a phase separation.

Winsor expressed qualitatively this variation in dispersing tendency by

$$R = \frac{A_{co}}{A_{cw}} \quad (3.3.4)$$

To account for the structure of the oil, and the interactions between surfactant molecules, an extended version of the original R-ratio was proposed [26]:

$$R = \frac{(A_{co} - A_{oo} - A_{LL})}{(A_{cw} - A_{ww} - A_{HH})} \qquad (3.3.5)$$

As mentioned before, in many cases, A_{Hco} and A_{Lcw} are negligible, so A_{co} and A_{cw} can be approximated respectively to A_{Lco} and A_{Hcw}.

In brief, Winsor's primary concept is that this R-ratio of cohesive energies, stemming from interaction of the interfacial layer with oil, divided by energies resulting from interactions with water, determines the preferred interfacial curvature. Thus, if $R > 1$, the interface tends to increase its area of contact with oil while decreasing its area of contact with water. Thus oil tends to become the continuous phase and the corresponding characteristic system is type II (Winsor II). Similarly, a balanced interfacial layer is represented by $R = 1$.

2. Packing parameter and microemulsion structures

Changes in film curvature and microemulsion type can be addressed quantitatively in terms of geometric requirements. This concept was introduced by Israelachivili et al. [27] and is widely used to relate surfactant molecular structure to interfacial topology. As described in Section 2.3.3, the preferred curvature is governed by relative areas of the head group, a_o, and the tail group, V/l_c (see Figure 2.6 for the possible aggregate structures). In terms of microemulsion type,

- if $a_o > V/l_c$, then an oil-in-water microemulsion forms,
- if $a_o < V/l_c$, then an water-in-oil microemulsion forms,
- if $a_o = V/l_c$, then a middle-phase microemulsion is the preferred structure.

3. Hydrophilic-Lipophilic Balance (HLB)

Another concept relating molecular structure to interfacial packing and film curvature is HLB, the hydrophilic-lipophilic balance. It is generally expressed as an empirical equation based on the relative proportions of hydrophobic and hydrophilic groups within the molecule. The concept was first introduced by Griffin [28] who characterized a number of surfactants, and derived an empirical equation for non-ionic alkyl polyglycol ethers (C_iE_j) based on the surfactant chemical composition [29]:

$$\text{HLB} = (E_j \text{wt\%} + \text{OH wt\%})/5 \qquad (3.3.6)$$

where E_i wt% and OH wt% are the weight percent of ethylene oxide and hydroxide groups respectively.

Davies *et al.* [30] proposed a more general empirical equation that associates a constant to the different hydrophilic and hydrophobic groups:

$$\text{HLB} = [(n_H \times H) - (n_L \times L)] + 7 \qquad (3.3.7)$$

where H and L are constants assigned to hydrophilic and hydrophobic groups respectively, and n_H and n_L the number of these groups per surfactant molecule.

For bicontinuous structures, i.e., zero curvature, it was shown that HLB $\approx$ 10 [31]. Then w/o microemulsions form when HLB < 10, and O/W microemulsion when HLB > 10. HLB and packing parameter describe the same basic concept, though the latter is more suitable for microemulsions. The influence of surfactant geometry and system conditions on HLB numbers and packing parameter is illustrated in Figure 3.4.

4. Phase Inversion Temperature (PIT)

Non-ionic surfactants form water-oil microemulsions (and emulsions) with a high temperature sensitivity. In particular, there is a specific phase inversion temperature (PIT) and the film curvature changes from positive to negative. This critical point was defined by Shinoda *et al.* [32]:

- if T < PIT, an oil-in-water microemulsion forms (Winsor I),
- if T > PIT, a water-in-oil microemulsion forms (Winsor II),
- at T = PIT, a middle-phase microemulsion exists (Winsor III) with a spontaneous curvature equal to zero, and a HLB number (Eq. 3.3.6) approximately equal to 10.

The HLB number and PIT are therefore connected; hence the term HLB temperature is sometimes employed [33].

3.3.2 Surfactant film properties

An alternative, more physically realistic, approach is to consider mechanical properties of a surfactant film at an oil-water interface. This film can be characterized by three phenomenological constants: tension, bending rigidity, and spontaneous curvature. Their relative importance depends on the constraints felt by the film. It is important to understand how these parameters relate to interfacial stability since surfactant films determine

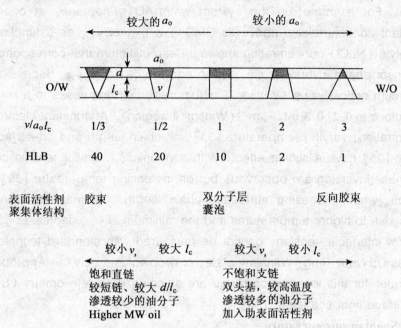

Figure 3.4 Effect of molecular geometry and system conditions on the packing parameter and HLB number (after Israelachvili [31]).

the static and dynamic properties of microemulsions (and emulsions). These include phase behaviour and stability, structure, and solubilisation capacity.

1. Ultra-low interfacial tension

Interfacial (or surface) tensions, γ, were defined in Chapter 2 for planar surfaces, and the same principle applies for curved liquid-liquid interfaces, i.e., it corresponds to the work required to increase interfacial area by unit amount. As mentioned in Section 3.2.1, microemulsion formation is accompanied by ultra-low interfacial oil-water tensions, $\gamma_{o/w}$, typically 10^{-2} to 10^{-4} mN·m^{-1}. They are affected by the presence of a co-surfactant, as well as electrolyte and/or temperature, pressure, and oil chain length. Several studies have been reported on the effect of such variables on $\gamma_{o/w}$. In particular, Aveyard and coworkers performed several systematic interfacial tension measurements on both ionics [34, 35] and non-ionics [36], varying oil chain length, temperature, and electrolyte content. For example, in the system water-AOT-n-heptane, at constant surfactant concentration (above its CMC), a plot of $\gamma_{o/w}$ as a function of electrolyte (NaCl) concentration shows a deep minimum that corresponds to the Winsor phase inversion; i. e. , upon addition of NaCl, $\gamma_{o/w}$ decreases to a minimum critical value (Winsor III structure), then increases to a limiting value close to 0. 2- 0. 3 mN · m^{-1}(Winsor II region). At constant electrolyte concentration, varying temperature [34], oil chain length and co-surfactant content [35] have a similar effect. With non-ionics, a similar tension curve and phase inversion are observed, but on increasing temperature [36]. In addition, when increasing surfactant chain length, the interfacial tension curves shift to higher temperatures and the minimum in $\gamma_{o/w}$ decreases [37]. Ultra-low interfacial tensions cannot be measured with standard techniques such as Du Nouy Ring, Wilhelmy plate, or drop volume (DVT). Appropriate techniques for this low tension range are spinning drop tensiometry (SDT) and surface light scattering [38].

2. Spontaneous curvature

Spontaneous (or natural or preferred) curvature C_o is defined as the curvature formed by a surfactant film when a system consists of equal amounts of water and oil. Then, there is no constraint on the film, which is free to adopt the lowest free energy state. Whenever one phase is predominant, there is a deviation from C_o. In principle, every point on a surface possesses two principal radii of curvature, R_1 and R_2 and their associated principal curvatures are $C_1 = 1/R_1$ and $C_2 = 1/R_2$. Mean and Gaussian curvatures are used to define the bending of surfaces. They are defined as follows [39]:

Mean curvature: $C = 1/2 \ (1/R_1 + 1/R_2)$

Gaussian curvature: $\kappa = 1/R_1 \times 1/R_2$

C_1 and C_2 are determined as follows: every point on the surface of the surfactant film has two principal radii of curvature, R_1 and R_2 as shown in Figure 3. 5. If a circle is placed tangentially to a point p on the surface and if the circle radius is chosen so that its second derivative at the contact point equals that of the surface in the direction of the tangent (of normal vector, n), then the radius of the circle is a radius of curvature of the surface. The curvature of the surface is described by two such circles chosen in orthogonal (principal) directions as shown in Figure 3.5(a).

For a sphere, R_1 and R_2 are equal and positive (Figure 3.5(b)). For a cylinder R_2 is indefinite (Figure 3.5(c)) and for a plane, both R_1 and R_2 are indefinite. In the special case of a saddle, $R_1 = -R_2$, i.e., at every point the surface is both concave and convex (Figure 3.5(d)). Both a plane and saddle have the property of zero mean curvature.

The curvature C_o depends both on the composition of the phases it separates and on surfactant type. One argument applied to the apolar side of the interface is that oil can penetrate to some extent between the surfactant hydrocarbon tails. The more extensive the penetration, the more curvature is imposed toward the polar side. This results in a decrease of C_o since, by convention, positive curvature is toward oil (and negative toward water). The longer the oil chains, the less they penetrate the surfactant film and the smaller the effect on C_o. Recently, Eastoe *et al.* have studied the extent of solvent penetration in microemulsions stabilized by di-chained surfactants, using SANS and selective deuteration. Results suggested that oil penetration is a subtle effect, which depends on the chemical structures of both surfactant and oil. In particular, unequal surfactant chain length [40-43] or presence of C=C bonds [44] result in a more disordered surfactant/oil interface, thereby providing a region of enhanced oil mixing. For symmetric di-chained surfactants (e.g., DDAB and AOT), however, no evidence for oil mixing was found [42]. The effect of alkane structure, and molecular volume on the oil penetration was also investigated with n-heptane, and cyclohexane. The results indicate that heptane is essentially absent from the layers, but the more compact cyclohexane has a greater penetrating effect [43].

Surfactant type, and nature of the polar head group, also influences C_o through different interactions with the polar (aqueous) phase:
- For ionic surfactants electrolyte content and temperature affect the spontaneous curvature in opposite ways. An increase in salt concentration screens electrostatic head group repulsions—i.e.,

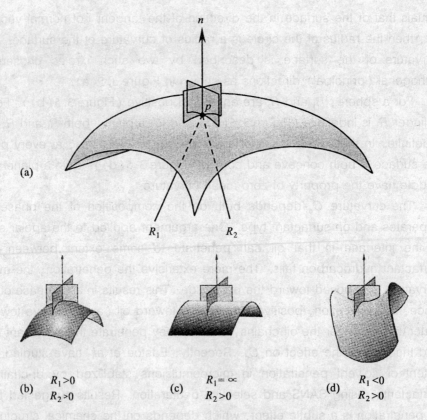

Figures 3.5 Principal curvatures of different surfaces. (a) Intersection of the surfactant film surface with planes containing the normal vector (n) to the surface at the point p. (b) convex curvature, (c) cylindrical curvature, (d) saddle-shaped curvature. After Hyde et al. [39].

decreases head group area—so the film curves more easily toward water, leading to a decrease in C_o. Raising temperature has two effects: (1) an increase in electrostatic repulsions between head groups due to higher counterion dissociation, so C_o increases; (2) more gauche conformations are induced in the surfactant chains, which become more coiled, resulting in a decrease in C_o. Therefore the combined effects of temperature on the apolar chains and on electrostatic interactions are competitive. The electrostatic term is believed to be slightly dominant, so C_o increases weakly with increasing temperature.

- For non-ionic surfactants, unsurprisingly, electrolytes have very little effect on C_o, whereas temperature is a critical parameter due to the strong dependence of their solubility (in water or oil) on temperature. For surfactants of the C_iE_j type as temperature increases water becomes a less good solvent for the hydrophilic units and penetrates less into the surfactant layer. In addition, on the other side of the film, oil can penetrate further into the hydrocarbon chains, so that increasing temperature for this type of surfactant causes a strong decrease in C_o. This phenomenon explains the strong temperature effects on the phase equilibria of such surfactants as shown in Figure 2.8 (see Chapter 2).

Thus, by changing external parameters such as temperature, nature of the oil or electrolyte concentration, the spontaneous curvature can be tuned to the appropriate value, and so drive transitions between Winsor systems. Other factors affect C_o in a similar fashion; they include varying the polar head group, type and valency of counterions, length and number of apolar chains, adding a co-surfactant, or mixing surfactants.

3. Film bending rigidity

Film rigidity is an important parameter associated with interfacial curvature. The concept of film bending energy was first introduced by Helfrich [45] and is now considered as an essential model for understanding microemulsion properties. It can be described by two elastic moduli [46] that measure the energy required to deform the interfacial film from a preferred mean curvature:

- the mean bending elasticity (or rigidity), K, associated with the mean curvature, that represents the energy required to bend unit area of surface by unit amount. K is positive, i.e., spontaneous curvature is favoured;

- the elasticity $\bar{K}$, associated with Gaussian curvature, and hence accounts for film topology. $\bar{K}$ is negative for spherical structures or positive for bicontinuous cubic phases.

Theoretically, it is expected that bending moduli should depend on surfactant chain length [47], area per surfactant molecule in the film [48] and electrostatic head group interactions [49].

The film rigidity theory is based on the interfacial free energy associated with film curvature. The free energy, F, of a surfactant layer at a liquid interface may be given by the sum of an interfacial energy term, F_i, a bending energy term, F_b, and an entropic term, F_{ent}. For a droplet type structure this is written as [50]

$$F = F_i + F_b + F_{ent} = \gamma A + \int\!\!\int \left[\frac{K}{2}(C_1 + C_2 - 2C_0)^2 + \bar{k}\,C_1 C_2 \right] dA + n K_B T f(\phi) \qquad (3.3.8)$$

where γ is the interfacial tension, A is the total surface area of the film, K is the mean elastic bending modulus, $\bar{K}$ is the Gaussian bending modulus, C_1 and C_2 are the two principal curvatures, C_0 the spontaneous curvature, n is the number of droplets, k_B is the Boltzmann constant, and $f(\phi)$ is a function accounting for the entropy of mixing of the microemulsion droplets, where ϕ is the droplet core volume fraction. For dilute systems where $\phi < 0.1$, it was shown that $f(\phi) = [\ln(\phi) - 1]$ [50]. Microemulsion formation is associated with ultra-low interfacial tension, γ so the γA term is small compared to F_b and F_{ent} and can be ignored as an approximation.

As mentioned previously, the curvatures C_1, C_2 and C_0 can be expressed in terms of radii as $1/R_1$, $1/R_2$, and $1/R_0$ respectively. For spherical droplets, $R_1 = R_2 = R$, and the interfacial area is $A = n 4\pi R^2$. Note that R and R_0 are core radii rather than droplet radii [50]. Solving Eq. 3.3.8 and dividing by area A, the total free energy, F, for spherical droplets (of radius R) is expressed as

$$\frac{F}{A} = 2K\left(\frac{1}{R} - \frac{1}{R_0}\right)^2 + \frac{\bar{K}}{R^2} + \left[\frac{k_B T}{4\pi R^2} f(\phi)\right] \qquad (3.3.9)$$

For systems where the solubilisation boundary is reached (W I or W II region), a microemulsion is in equilibrium with an excess phase of the solubilisate and the droplets have achieved their maximum size, i.e., the maximum core radius, R_{max}^{av}. Under this condition the minimization of the total free energy leads to a relation between the spontaneous radius, R_0, and the elastic constants K and $\bar{K}$ [51]:

$$\frac{R_{max}^{av}}{R_o} = \frac{2K + \bar{K}}{2K} + \frac{k_B T}{8\pi K} f(\phi) \qquad (3.3.10)$$

A number of techniques have been used to determine K and $\bar{K}$ separately, in particular, ellipsometry, X-ray reflectivity, and small-angle X-ray scattering (SAXS) techniques [52-54]. De Gennes and Taupin [55] have developed a model for bicontinuous microemulsions. For $C_o = 0$ the layer is supposed to be flat in the absence of thermal fluctuations. They introduced the term ξ_K, the persistence length of the surfactant layer that relates to K via:

$$\xi_K = a \exp(2\pi K / k_B T) \qquad (3.3.11)$$

where a is a molecular length and ξ_K is the correlation length for the normals to the layer, i.e., the distance over which this layer remains flat in the presence of thermal fluctuations. ξ_K is extremely sensitive to the magnitude of K. When $K \gg k_B T$, ξ_K is macroscopic, i.e., the surfactant layer is flat over large distances and ordered structures such as lamellar phases may form. If K is reduced to $\sim k_B T$ then ξ_K is microscopic, ordered structures are unstable and disordered phases such as microemulsions may form. Experiments reveal that K is typically between 100 $k_B T$ for condensed insoluble monolayers [56] and about 10 $k_B T$ for lipid bilayers [57-59] but can decrease below $k_B T$ in microemulsion systems [60]. The role of $\bar{K}$ is also important, however, there are few measurements of this quantity in the literature [e.g., 53, 61]. Its importance in determining the structure of surfactant-oil-water mixtures is still far from clear.

An alternative, more accessible, method to quantify film rigidities is to calculate the composite parameter $(2K + \bar{K})$ using tensiometry and SANS techniques. This parameter can be derived for droplet microemulsion at the solubilisation boundary, W I or W II system, by combining the radius of the droplet with interfacial tensions or droplet polydispersity. Two expressions can be derived from Eq. 3.3.9 and 3.3.10.

(1) *Using the interfacial tension $\gamma_{o/w}$ (measured by SLS or SDT) and the maximum mean core radius R_{max}^{av} (measured by SANS see Chapter 4)*:
$\gamma_{o/w}$ at the interface between microemulsion and excess phases at the

solubilisation boundary can be expressed in terms of these elastic moduli and R_{max}^{av} [52]. Any new area created must be covered by a monolayer of surfactant, and so this energy may be calculated in the case of W I or W II systems since the surfactant monolayer is taken from around the curved microemulsion droplets [56]. To do this it is necessary to unbend the surfactant film, introducing a contribution from K, of $2K/(R_{max}^{av})^2$. The resulting change in the number of microemulsion droplets introduces an entropic contribution and a contribution due to the change in topology involving $\bar{K}$, of $\bar{K}/(R_{max}^{av})^2$. So the interfacial tension between the microemulsion and excess phase is given by

$$\gamma_{o/w} = \frac{2K + \bar{K}}{(R_{max}^{av})^2} + \frac{k_B T}{4\pi (R_{max}^{av})^2} f(\phi) \quad (3.3.12)$$

which gives for the bending moduli:

$$2K + \bar{K} = \gamma_{o/w}(R_{max}^{av})^2 - \frac{k_B T}{4\pi} f(\phi) \quad (3.3.13)$$

(2) Using the Schultz polydispersity width $p = \sigma/R_{max}^{av}$ obtained from SANS analysis:

Droplet polydispersity relates to the bending moduli through thermal fluctuations of the microemulsion droplets. Safran [62] and Milner [63] described the thermal fluctuations by an expansion of the droplet deformation in terms of spherical harmonics. The principal contribution to these fluctuations was found to arise from the deformation mode $l = 0$ only [50]; and $l = 0$ deformations are fluctuations in droplet size, i.e., changes of the mean droplet radius and hence the droplet polydispersity. In the case of the two phase equilibria at maximum solubilisation (W I or W II), this polydispersity, p, may be expressed as a function of K and $\bar{K}$:

$$p^2 = \frac{u_o^2}{4\pi} = \frac{k_B T}{8\pi(2K + \bar{K}) + 2k_B T f(\phi)} \quad (3.3.14)$$

where u_o is the fluctuation amplitude for the $l = 0$ mode. This polydispersity is given by the SANS Schultz polydispersity parameter σ/R_{max}^{av} [64], and Eq. 3.3.14 can be written:

$$2K + \bar{K} = \frac{k_B T}{8\pi(\sigma/R_{max}^{av})^2} - \frac{k_B T}{4\pi} f(\phi) \quad (3.3.15)$$

Therefore 3.3.12 and Eq. 3.3.15 give two accessible expressions for the sum $(2K + \bar{K})$ using data from SANS and tensiometry. This approach has been shown to work well with non-ionic films in W I systems [50, 65], and also cationic [64] and zwitterionic [66] layers in W II microemulsions. Figure 3.6 shows results for these latter two classes of system, as a function of surfactant alkyl carbon number $n - C$. The good agreement between equation. 3.3.12 and 3.3.14 suggests they can be used with confidence. These values are in line with current statistical mechanical theories [48], which suggest that K should vary as $n - C^{2.5}$ to $n - C^3$, whereas there is only a small effect on $\bar{K}$.

3.3.3 Phase behaviour

Solubilisation and interfacial properties of microemulsions depend upon pressure, temperature and also on the nature and concentration of the components. The determination of phase stability diagrams (or phase maps), and location of the different structures formed within these water (salt)-oil-surfactant-alcohol systems in terms of variables are, therefore, very important. Several types of phase diagram can be identified depending on the number of variables involved. In using an adequate mode of representation, it is possible to describe not only the limits of existence of the single and multiphase regions, but also to characterize equilibria between phases (tie-lines, tie-triangles, critical points, etc.). Below is a brief description of ternary and binary phase maps, as well as the phase rule that dictates their construction.

1. Phase rule

The phase rule enables the identification of the number of variables (or degrees of freedom) depending on the system composition and conditions. It is generally written as [67]

$$F = C - P + 2 \quad (3.3.16)$$

where F is the number of possible independent changes of state or degrees of freedom, C the number of independent chemical constituents, and P the number of phases present in the system. A system is called invariant, monovariant, bivariant, and so on, according to whether F is zero, 1, 2,

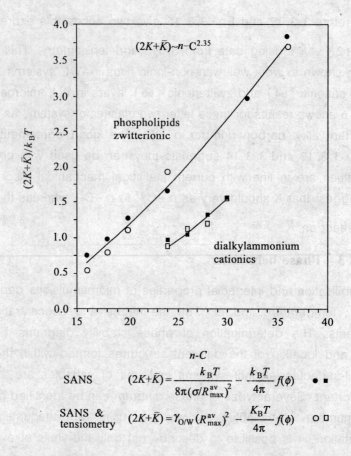

Figure 3.6 Film rigidities ($2K + \bar{K}$) as a function of total alkyl carbon number n-C from Winsor II microemulsions. The lines are guides to the eye. After Eastoe et al. [64,66].

and so on. For example, in the simplest case of a system composed of three components and two phases, F is univariant at a fixed temperature and pressure. This means that the mole or weight fraction of one component in one of the phases can be specified but all other compositions in both phases are fixed. In general, microemulsions contain at least three components: oil (O), water (W) and amphiphile (S), and as mentioned previously a co-surfactant (alcohol) and/or an electrolyte are usually added to tune the system stability. These can be considered as simple O-W-S systems; whenever a co-surfactant is used, the ratio oil: alcohol is kept constant and it is assumed that the alcohol does not interact with any other component so that the mixture can be treated (to a first approximation) as a three-component system. At constant pressure, the composition—temperature phase behaviour can be presented in terms of a phase prism, as illustrated in Figure 3.7. However, the construction of such a phase map is rather complex and time consuming so it is often convenient to simplify the system by studying specific phase-cuts. The number of variables can be reduced either by keeping one term constant and/or by combining two or more variables. Then, ternary and binary phase diagrams are produced.

2. Ternary phase diagrams

At constant temperature and pressure, the ternary phase diagram of a simple three-component microemulsion is divided into two or four regions as shown in Figure 3.8. In each case, every composition point within the single-phase region above the demixing line corresponds to a microemulsion. Composition points below this line correspond to multiphase regions comprising in general microemulsions in equilibrium with either an aqueous or an organic phase or both, i.e., Winsor type systems (see Section 3.3.1).

Any system whose overall composition lies within the two-phase region (e.g., point o in Figures 3.8(a) and 3.8(c)) will exist as two phases whose compositions are represented by the ends of the "tie-line", i.e., a segment formed by phases m and n. Therefore, every point on a particular tie-line has identical coexisting phases (m and n) but of different relative volumes. When the two conjugate phases have the same composition ($m = n$), this corresponds to the plait (or critical) point, p.

If three phases coexist (Figure 3.8(b)), i.e., corresponding to W III, the system at constant temperature and pressure is, according to the phase rule, invariant. Then, there is a region of the ternary diagram that consists of three-phase systems having invariant compositions and whose boundaries are tie-lines in the adjacent two-phase regions that surround it. This region of three-phase invariant compositions is therefore triangular in form and called "tie-triangle" [26]. Any overall composition, such as point q (Figure 3.8(b)) lying within the tie-triangle will divide into three phases having compositions corresponding to the vertices A, B, and C of the triangle. The compositions A, B, and C are invariant in the sense that varying the position

q, the overall composition, throughout the triangle will result in variations in the amounts of the phases A, B, and C but not in their composition.

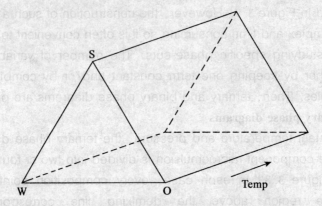

Figure 3.7 The phase prism, describing the phase behaviour of a ternary system at constant pressure.

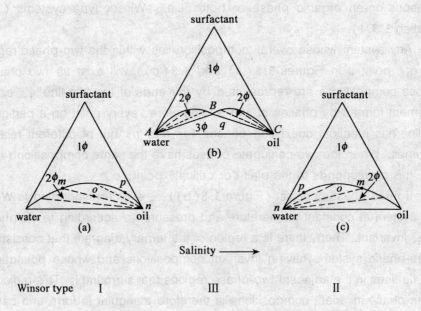

Figure 3.8 Ternary diagram representations of two-and three-phase regions formed by simple water-oil-surfactant systems at constant temperature and pressure. (a) Winsor I type, (b) Winsor II type, (c) Winsor III type systems.

Binary phase diagrams

As mentioned previously, ternary diagrams can be further simplified by fixing some parameters and/or combining two variables together (e.g., water and electrolyte into brine, or water and oil into water-to-oil ratio), i.e., reducing the degrees of freedom. Then, determining the phase diagram of such systems reduces to a study of a planar section through the phase prism. Examples of such pseudo-binary diagrams are given in Figures 3.9 to 3.11 for non-ionic and anionic surfactants.

Figure 3.9 shows the schematic phase diagram for a non-ionic surfactant-water-oil ternary system. Since temperature is a crucial variable in the case of non-ionics, the pseudo-binary diagram is represented by the planar section defined by $\phi_w = \phi_o$, where ϕ_w and ϕ_o are the volume fractions of water and oil respectively. Then, at constant pressure, defining the system in a single-phase region requires the identification of two independent variables ($F=2$), i.e., temperature and surfactant concentration. The section shown in Figure 3.9(b) can be used to determine T_L and T_U, the lower and upper temperatures, respectively, of the phase equilibrium W+M+O (with M, the microemulsion phase), and the minimum amount of surfactant necessary to solubilise equal amounts of water and oil, denoted C_s^* [68]. The lower C_s^* the more efficient the surfactant. Figure 3.10 illustrates the determination of a second possible section for a non-ionic surfactant-water-oil ternary system: pressure and surfactant concentration are kept constant, leaving the two variables, temperature and water-to-oil ratio ($\phi_{w\text{-}o}$). This diagram shows the various surfactant phases obtained as a function of temperature and water-to-oil ratio [68]. The third example (Figure 3.11) concerns an anionic surfactant, Aerosol-OT. In order to obtain $F=2$ when defining the ternary W-O-S system in a single-phase region at constant pressure, the surfactant concentration parameter is fixed. Then, the two variables are temperature and w, the water-to-surfactant molar ratio defined as $w = [\text{water}] / [\text{surfactant}]$. w represents the number of water molecules solubilised per surfactant molecule, so that this phase diagram characterizes the surfactant efficiency, as a microemulsifier.

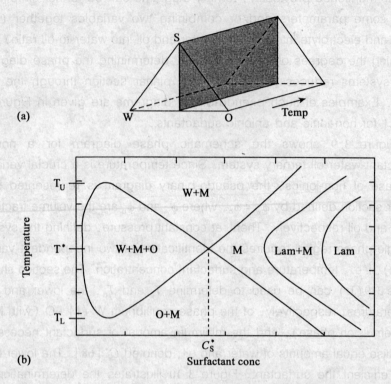

Figure 3.9 Binary phase behaviour in ternary microemulsion systems formed with non-ionic surfactants. (a) Illustration of the section through the phase prism at equal water and oil content. (b) Schematic phase diagram plotted as temperature versus surfactant concentration C_s. T_L and T_U are the lower and upper temperatures, respectively, of the phase equilibrium W + M + O. T^* is the temperature at which the three-phase triangle is an isosceles, i.e., when the middle-phase microemulsion contains equal amounts of water and oil. This condition is also termed 'balanced'. C_s^* is the surfactant concentration in the middle-phase microemulsion at balanced conditions. 'Lam' denotes a lamellar liquid crystalline phase. After Olsson and Wennerström [68].

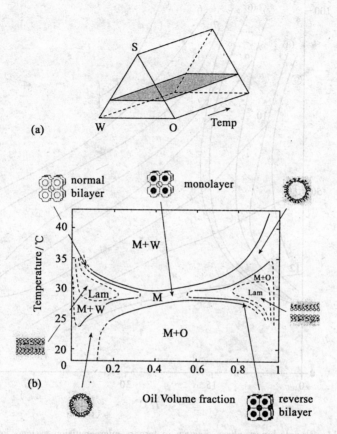

Figure 3.10 Binary phase behaviour in ternary microemulsion systems formed with non-ionic surfactants. (a) Illustration of a section at constant surfactant concentration through the phase prism. (b) Schematic phase diagram, plotted as temperature versus volume fraction of oil, ϕ_o, at constant surfactant concentration. Also shown are various microstructures found in different regions of the microemulsion phase, M. At higher temperatures the liquid phase is in equilibrium with excess water (M + W), and at lower temperatures with excess oil (M + O). At intermediate temperatures a lamellar phase is stable at higher water contents and higher oil contents, respectively. After Olsson and Wennerström [68].

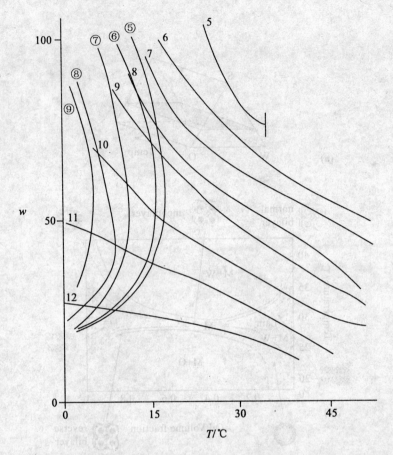

Figure 3.11 Pseudo-binary phase diagram in ternary microemulsion systems formed with the anionic surfactant Aerosol-OT (AOT) in various straight-chain alkane solvents. The wate-to-surfactant molar ratio, w, is plotted versus temperature at constant surfactant concentration and pressure. Alkane carbon numbers are indicated; ringed numbers correspond to the lower temperature (solubilisation) boundary, T_L, and un-ringed numbers to the upper temperature (haze) boundary, T_U. The single phase microemulsion region is located between T_L and T_U. Below T_L the system consists of a microemulsion phase in equilibrium with excess water (W II type), and above T_U the single microemulsion phase separates into a surfactant-rich phase and an oil phase. After Fletcher et al.[16].

REFERENCES

1. Danielsson, I.; Lindman, B. *Colloids Surf.* A 1981, *3*, 391.

2. Sjöblom, J. ; Lindberg, R. ; Friberg, S. E. *Adv. Colloid Interface Sci.* 1996, 125.
3. Schulman, J. H. ; Stoeckenius, W. ; Prince, M. *J. Phys. Chem.* 1959, 63, 1677.
4. Shinoda, K. ; Friberg, S. *Adv. Colloid Interface Sci.* 1975, 4, 281.
5. Adamson, A. W. *J. Colloid Interface Sci.* 1969, 29, 261. 3.
6. Friberg, S. E. ; Mandell, L. ; Larsson, M. *J. Colloid Interface Sci.* 1969, 29, 155.
7. Shah, D. O. , Ed. '*Surface Phenomena in Enhanced Recovery*' , Plenum Press, 1981, New York.
8. Overbeek, J. Th. G. *Faraday Discuss. Chem. Soc.* 1978, 65, 7.
9. Tadros, Th. F. ; Vincent, B. in ' *Encyclopaedia of Emulsion Technology*' Becher, P. Ed. , Vol. 1, Marcel Dekker, 1980, New York.
10. Kunieda, H. ; Shinoda, K. *J. Colloid Interface Sci.* 1980, 75, 601.
11. Chen, S. J. ; Evans, F. D. ; Ninham, B. W. *J. Phys. Chem.* 1984, 88, 1631.
12. Kahlweit, M. ; Strey, R. ; Busse, G. *J. Phys. Chem.* 1990, 94, 3881.
13. Hunter, R. J. '*Introduction to Modern Colloid Science*' , Oxford University Press, 1994, Oxford.
14. Lekkerkerker, H. N. W. ; Kegel, W. K. ; Overbeek, J. Th. G. *Ber. Bunsenges Phys. Chem.* 1996, 100, 206.
15. Ruckenstein, E. ; Chi, J. C. *J. Chem. Soc. Faraday Trans.* 1975, 71, 1690.
16. Fletcher, P. D. I. ; Howe, A. M. ; Robinson, B. H. *J. Chem. Soc. Faraday Trans.* 1 1987, 83, 985.
17. Fletcher, P. D. I. ; Clarke, S. ; Ye, X. *Langmuir* 1990, 6, 1301.
18. Biais, J. ; Bothorel, P. ; Clin, B. ; Lalanne, P. *J. Colloid Interface Sci.* 1981, 80, 136.
19. Friberg, S. ; Mandell, L. ; Larson, M. *J. Colloid Interface Sci.* 1969, 29, 155.
20. Fletcher, P. D. I. ; Horsup, D. I. *J. Chem. Soc. Faraday Trans.* 1 1992, 88, 855.

21. Winsor, P. A. *Trans. Faraday Soc.* 1948, *44*, 376.
22. Bellocq, A. M. ; Biais, J. ; Bothorel, P. ; Clin, B. ; Fourche, G. ; Lalanne, P. ; Lemaire, B. ; Lemanceau, B. ; Roux, D. *Adv. Colloid Interface Sci.* 1984, *20*, 167.
23. Bancroft, W. D. *J. Phys. Chem.* 1913, *17*, 501.
24. Clowes, G. H. A. *J. Phys. Chem.* 1916, *20*, 407.
25. Adamson, A. W. '*Physical Chemistry of Surfaces*', Interscience, 1960, p 393.
26. Bourrel, M. ; Schechter, R. S. '*Microemulsions and Related Systems*', Marcel Dekker, 1988, New York.
27. Israelachvili, J. N. ; Mitchell, D. J. ; Ninham, B. W. *J. Chem. Soc. Faraday Trans.* 2 1976, *72*, 1525.
28. Griffin, W. C. *J. Cosmetics Chemists* 1949, *1*, 311.
29. Griffin, W. C. *J. Cosmetics Chemists* 1954, *5*, 249.
30. Davies, J. T. *Proc. 2nd Int. Congr. Surface Act.* Vol. 1 Butterworths, 1959, London.
31. Israelachvili, J. N. *Colloids Surf. A* 1994, *91*, 1.
32. Shinoda, K. ; Saito, H. *J. Colloid Interface Sci.* 1969, *34*, 238.
33. Shinoda, K. ; Kunieda, H. in '*Encyclopaedia of Emulsion Technology*', Becher, P. Ed., Vol. 1, Marcel Dekker, 1983, New York.
34. Aveyard, R. ; Binks, B. P. ; Clarke, S. ; Mead, J. *J. Chem. Soc. Faraday Trans. 1* 1986, *82*, 125.
35. Aveyard, R. ; Binks, B. P. ; Mead, J. *J. Chem. Soc. Faraday Trans. 1* 1986, *82*, 1755.
36. Aveyard, R. ; Binks, B. P. ; Fletcher, P. D. I. *Langmuir* 1989, *5*, 1210.
37. Sottmann, T. ; Strey, R. *Ber. Bunsenges Phys. Chem.* 1996, *100*, 237.
38. Langevin, D., Ed. '*Light Scattering by Liquid Surfaces and Complementary Techniques*', Marcel Dekker, 1992, New York.
39. Hyde, S. ; Andersson, K. ; Larsson, K. ; Blum, Z. ; Landh, S. ; Ninham, B. W. '*The Language of Shape*', Elsevier, 1997, Amsterdam.

40. Eastoe, J. ; Dong, J. ; Hetherington, K. J. ; Steytler, D. C. ; Heenan, R. K. *J. Chem. Soc. Faraday Trans.* 1996, *92*, 65.
41. Eastoe, J. ; Hetherington, K. J. ; Sharpe, D. ; Dong, J. ; Heenan, R. K. ; Steytler, D. C. *Langmuir* 1996, *12*, 3876.
42. Eastoe, J. ; Hetherington, K. J. ; Sharpe, D. ; Dong, J. ; Heenan, R. K. ; Steytler, D. C. *Colloids Surf. A* 1997, *128*, 209.
43. Eastoe, J. ; Hetherington, K. J. ; Sharpe, D. ; Steytler, D. C. ; Egelhaaf, S. ; Heenan, R. K. *Langmuir* 1997, *13*, 2490.
44. Bumajdad, A. ; Eastoe, J. ; Heenan, R. K. ; Lu, J. R. ; Steytler, D. C. ; Egelhaaf, S. *J. Chem. Soc. Faraday Trans.* 1998, *94*, 2143.
45. Helfrich, W. *Z. Naturforsch.* 1973, *28c*, 693.
46. Kellay, H. ; Binks, B. P. ; Hendrikx, Y. ; Lee, L. T. ; Meunier, J. *Adv. Colloid Interface Sci.* 1994, *9*, 85.
47. Safran, S. A. ; Tlusty, T. *Ber. Bunsenges. Phys. Chem.* 1996, *100*, 252.
48. Szleifer, I. ; Kramer, D. ; Ben-Shaul, A. ; Gelbart, W. M. ; Safran, S. *J. Chem. Phys.* 1990, *92*, 6800.
49. Winterhalter, M. ; Helfrich, W. *J. Phys. Chem.* 1992, *96*, 327.
50. Gradzielski, M. ; Langevin, D. ; Farago, B. *Phys. Rev. E* 1996, *53*, 3900.
51. Safran, S. A. in '*Structure and Dynamics of Strongly Interacting Colloids and Supramolecular Aggregates in Solution*', Vol. 369 of *NATO Advanced Study Institute, Series C: Mathematical and Physical Sciences*, Chen, S. H. ; Huang, J. S. ; Tartaglia, P. Ed. ; Kluwer, Dortrecht, 1992.
52. Meunier, J. ; Lee, L. T. *Langmuir* 1991, *46*, 1855.
53. Kegel, W. K. ; Bodnar, I. ; Lekkerkerker, H. N. W. *J. Phys. Chem.* 1995, *99*, 3272.
54. Sicoli, F. ; Langevin, D. ; Lee, L. T. *J. Chem. Phys.* 1993, *99*, 4759.
55. De Gennes, P. G. ; Taupin, C. *J. Phys. Chem.* 1982, *86*, 2294.
56. Daillant, J. ; Bosio, L. ; Benattar, J. J. ; Meunier, J. *Europhys. Lett.* 1989, *8*, 453.
57. Shneider, M. B. ; Jenkins, J. T. ; Webb, W. W. *Biophys. J.*

1984, *45*, 891.
58. Engelhardt, H; Duwe, H. P. ; Sackmann, E. *J. Phys. Lett.* 1985, *46*, 395.
59. Bivas, I. ; Hanusse, P. ; Botherel, P. ; Lalanne, J. ; Aguerre-Chariol, O. *J. Physique* 1987, *48*, 855.
60. Di Meglio, J. M. ; Dvolaitzky, M. Taupin, C. *J. Phys. Chem.* 1985, *89*, 871.
61. Farago, B. ; Huang, J. S. ; Richter, D. ; Safran, S. A. ; Milner, S. T. *Progr. Colloid Polym. Sci.* 1990, *81*, 60.
62. Safran, S. A. *J. Chem. Phys.* 1983, *78*, 2073.
63. Milner, S. T. ; Safran, S. A. *Phys. Rev. A* 1987, *36*, 4371.
64. Eastoe, J. ; Sharpe, D. ; Heenan, R. K. ; Egelhaaf, S. *J. Phys. Chem. B* 1997, *101*, 944.
65. Gradzielski, M. ; Langevin, D. *J. Mol. Struct.* 1996, *383*, 145.
66. Eastoe, J. ; Sharpe, D. *Langmuir* 1997, *13*, 3289.
67. Rock, P. A. '*Chemical Thermodynamics*', MacMillan, 1969, London.
68. Olsson, U. ; Wennerström, H. *Adv. Colloid Interface Sci.* 1994, *49*, 113.

4. Scattering techniques

The determination of molecular organisation within colloidal systems is an important aspect when studying relationships between physical properties and molecular structure. Scattering techniques provide the most obvious methods for obtaining quantitative information on size, shape and structure of colloidal particles, since they are based on interactions between incident radiations (e. g., light, X-ray or neutrons) and particles. The size range of micelles, microemulsions, or other colloidal dispersions is approximately 10-10^4 Å, so valuable information can be obtained if the incident wavelength, λ, falls within this range. Therefore, microemulsion droplets or micelles, in the order of 10^2 Å in size, are well characterized by X-ray (λ = 0.5 -2.3 Å) and neutrons (λ =0.1 -30 Å), while for larger colloidal particles, light scattering (λ = 4,000-8,000 Å), is best. In addition, considering the Bragg equation that defines the angle of diffraction θ of radiation of wavelength λ for a separation of lattice planes d:

$$\lambda = 2d\sin\theta \qquad (4.0.1)$$

it can be seen that nanometre-sized particles such as microemulsion droplets will scatter at small angles, so that small-angle neutron scattering (SANS) can be used to study such systems [1].

Although the first neutron reactors were built in the late 1940's and 1950's, literature for application of neutron scattering to condensed matter appeared only in the late 1970's. In the last twenty years, with the development of more powerful neutron production sites, and progress in the technology of large area detectors and high resolution spectrometers, SANS has become a more accessible technique and, in particular, has been used successfully to study micellisation, microemulsion and liquid crystal structures. SANS is thus a relatively recent technique but is now one of the most powerful tools to characterize molecular aggregates.

In the following sections a summary of neutron scattering theory and methods for SANS data analysis is given.

4.1 GENERAL BACKGROUND

4.1.1 Neutrons

A neutron is an uncharged (electrically neutral) subatomic particle with mass $m = 1.675 \times 10^{-27}$ kg (1,839 times that of the electron), spin 1/2, and magnetic moment -1.913 nuclear magnetons. Neutrons are stable when bound in an atomic nucleus, whilst having a mean lifetime of approximately 1000 seconds as a free particle. The neutron and the proton form nearly the entire mass of atomic nuclei, so they are both called nucleons. Neutrons are classified according to their wavelength and energy as "epithermal" for short wavelengths ($\lambda \sim 0.1$ Å), "thermal", and "cold" for long wavelengths ($\lambda \sim 10$ Å). The desired range of λ is obtained by moderation of the neutrons during their production, either in reactors or spallation sources.

Neutrons interact with matter through strong, weak, electromagnetic and gravitational interactions. However, it is their interactions via two of these forces—the short-range strong nuclear force and their magnitude moments—that make neutron scattering such a unique probe for condensed-matter research. The most important advantages of neutrons over other forms of radiation in the study of structure and dynamics on a microscopic level are summarised below:

- Neutrons are uncharged, which allows them to penetrate the bulk of materials. They interact via the strong nuclear force with the nuclei of the material under investigation.
- The neutron has a magnetic moment that couples to spatial variations of magnetization on the atomic scale. They are therefore ideally suited to the study of magnetic structures, and the fluctuations and excitations of spin systems.
- The energy and wavelength of neutrons may be matched, often simultaneously, to the energy and length scales appropriate for the structure and excitations in condensed matter. The wavelength, λ, is

dependent on the neutron velocity following the de Broglie relation:

$$\lambda = \frac{h}{mv} \qquad (4.1.1)$$

where h is Planck's constant (6.636×10^{-34} Js) and v the particle velocity.

The associated kinetic energy is

$$E = 1/2 mv^2 \text{ or } E = \frac{h^2}{2(m\lambda)^2} \qquad (4.1.2)$$

Because their energy and wavelength depend on their velocity, it is possible to select a specific neutron wavelength by the time-of-flight technique.

- Neutron do not significantly perturb the system under investigation, so the results of neutron scattering experiments can be clearly interpreted.
- Neutrons are non-destructive, even to delicate biological materials.
- The high-penetrating power of neutrons allows probing the bulk of materials and facilitates the use of complex sample-environment equipment (e.g., for creating extremes of pressure, temperature shear and magnetic fields).
- Neutrons scatter from materials by interacting with the nucleus of an atom rather than the electron cloud. This means that the scattering power (cross-section) of an atom is not strongly related to its atomic number, unlike X-rays and electrons where the scattering power increases in proportion to the atomic number. Therefore, with neutrons light atoms such as hydrogen (deuterium) can be distinguished in the presence of heavier ones. Similarly, neighbouring elements in the periodic table generally have substantially different scattering cross sections and so can be distinguished. The nuclear dependence of scattering also allows isotopes of the same element to have substantially different scattering lengths for neutrons. Hence isotopic substitution can be used to label different parts of the molecules making up a material.

4.1.2 Neutron sources

Neutron beams may be produced in two general ways: by nuclear fission in reactor-based neutron sources, or by spallation in accelerator-based neutron sources. A brief description of these processes is given below, with particular reference to the two world's most intense neutron sources, i.e., the Institut Laue-Langevin (ILL) in Grenoble, France [2], and the ISIS Facility at the Rutherford Appleton Laboratory in Didcot, U.K. [3].

- *Reactor-based neutron source*: neutrons have traditionally been produced by fission in nuclear reactors optimised for high neutron brightness. In this process, thermal neutrons are absorbed by uranium-235 nuclei, which split into fission fragments and evaporate a very high-energy (MeV) constant neutron flux (hence the term "steady-state" or "continuous" source). After the high-energy (MeV) neutrons have been thermalised to MeV energies in the surrounding moderator, beams are emitted with a broad band of wavelengths. The energy distribution of the neutrons can be shifted to higher energy (shorter wavelength) by allowing them to come into thermal equilibrium with a "hot source" (at the ILL this is a self-heating graphite block at 2,400 K), or to lower energies with a "cold source" such as liquid deuterium at 25 K [4]. The resulting Maxwell distributions of energies have the characteristic temperatures of the moderators (Figure 4.1(a)). Wavelength selection is generally achieved by Bragg scattering from a crystal monochromator or by velocity selection through a mechanical chopper. In this way high-quality, high-flux neutron beams with a narrow wavelength distribution are made available for scattering experiments. The most powerful of the reactor neutron sources in the world today is the 58 MW HFR (High-Flux Reactor) at the ILL.

- *Accelerator-based pulsed neutron source*: in these sources neutrons are released by bombarding a heavy-metal target (e.g., U, Ta, W), with high-energy particles (e.g., H^+) from a high-power accelerator—a process known as spallation. The methods of particles acceleration tend to produce short intense bursts of high-energy protons, and hence pulses of neutrons. Spallation releases much less heat per useful neutron than fission (typically 30 MeV per neutron, compared with 190 MeV in fission). The low heat dissipation means that pulsed sources can deliver high neutron brightness-exceeding that of the most advanced steady-state sources-with significantly less heat generation in the target. The most powerful spallation neutron source in the world is the ISIS facility. It is based around a 200 μA, 800 MeV, proton synchrotron operating at 50 Hz, and a tantalum (Ta) target which releases approximately 12 neutrons for every incident proton.

At ISIS, the production of particles energetic enough to result in efficient spallation involves three stages (see Figure 4.2):

(1) Production of H^- ions (proton with two electrons) from hydrogen gas and acceleration in a pre-injector column to reach an energy of 665 keV.

(2) Acceleration of the H^- ions to 70 MeV in the linear accelerator (Linac) which consists of four accelerating tanks.

(3) Final acceleration in the synchrotron—a circular accelerator 52 m in diameter that accelerates 2.8×10^{13} protons per pulse to 800 MeV. As they enter the synchrotron, the H^- ions pass through a very thin (0.3 μm) alumina foil so that both electrons from each H^- ion are removed to produce a proton beam. After travelling around the synchrotron (approximately 10,000 revolutions), with acceleration on each revolution from electromagnetic fields, the proton beam of 800 MeV is kicked out of the synchrotron towards the neutron production target. The entire acceleration process is repeated 50 times a second.

Collisions between the proton beam and the target atom nuclei generate neutrons in large quantities and of very high energies. As in fission, they must be slowed by passage through moderating materials so that they have the right energy (wavelength) to be useful for scientific investigations. This is achieved by hydrogenous moderators around the target. These exploit the large inelastic-scattering cross-section of hydrogen to slow down the neutrons passing through, by repeated collisions with the hydrogen nuclei. The moderator temperature determines the spectral distributions of neutrons produced, and this can be tailored for different type of experiments (Figure

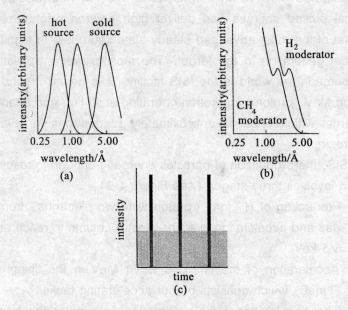

Figure 4.1 (a) Typical wavelength distributions for neutrons from a reactor, showing the spectra from a hot source (2,400 K), a thermal source and a cold source (25 K). The spectra are normalised so that the peaks of the Maxwell distributions are unity.
(b) Typical wavelength spectra from a pulsed spallation source. The H_2 and CH_4 moderators are at 20 K and 100 K respectively. The spectra have a high-energy "slowing" component and a thermalised component with a Maxwell distribution. Again the spectra are normalised at unity.
(c) Neutron flux as a function of time at a steady-state source (grey) and a pulsed source (black). Steady-state sources, such as ILL, have high time-averaged fluxes, whereas pulsed sources, such as ISIS, are optimised for high brightness (not drawn to scale). After [3].

4.1 (b)). The moderators at ISIS are ambient temperature water (316 K, H_2O), liquid methane (100 K, CH_4) and liquid hydrogen (20 K, H_2).

The characteristics of the neutrons produced by a pulsed source are therefore significantly different from those produced at a reactor (Figure 4.1 (c)). The time-averaged flux (in neutrons per second per unit area) of even the most powerful pulsed source is low in comparison with reactor sources. However, judicious use of time-of-flight (TOF) techniques that exploit the high brightness in the pulse can compensate for this. Using TOF

4. Scattering techniques

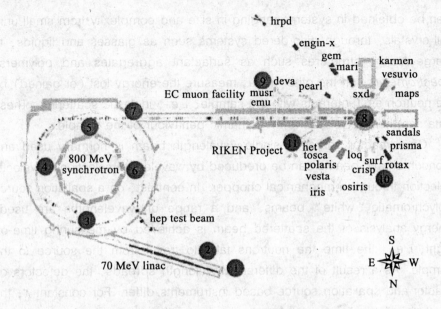

1. Ion source and per-injector
2. 70 MeV linear accelerator
3. 800 MeV synchrotron injection area
4. Fast kicker proton beam extraction
5. Synchrotron south side
6. Synchrotron west side
7. Extracted proton beam tunnel
8. ISIS target station
9. Experimental hall, south side
10. Experimental hall, north side
11. RIKEN superconducting pion decay line

Figure 4.2 Schematic layout of the spallation pulsed neutron source at the Rutherford Appleton Laboratory, ISIS, Didcot, U.K. Beam tubes radiate out from the ISIS target and deliver pulses of "white" neutrons—i.e., neutrons having a wide range of energies—to 18 instruments [3].

techniques on the white neutron beam gives a direct determination of the energy and wavelength of each neutron.

4.1.3 SANS instruments

In neutron scattering experiments, instruments count the number of scattered neutrons as a function of wave vector Q, which depends on the scattering angle θ and wavelength λ (see Section 4.4). For elastic scattering — i.e., when scattered neutrons have essentially identical energy to the incident neutrons—this corresponds to measuring with *diffractometers* the momentum change. Information about the spatial distribution of nuclei can then be obtained in systems ranging in size and complexity from small unit-cell crystals, through disordered systems such as glasses and liquids, to "large-scale" structures such as surfactant aggregates and polymers. *Spectrometers*, on the other hand, measure the energy lost (or gained) by the neutron as it interacts with the sample, i. e., inelastic scattering. These data can then be related to the dynamic behaviour of the sample.

On a reactor source a single-wavelength beam is normally used and monochromatic beams can be produced by wavelength selection by velocity selection through a mechanical chopper. In contrast, on a spallation source polychromatic "white" beams, and a range of wavelengths are used. Energy analysis of the scattered beam is achieved by measuring time-of-flight, i. e., the time the neutrons take to travel from the source to the sample. As a result of the different wavelength spreads, the detectors on reactor and spallation source based instruments differ. For constant λ, the scattering intensity must be measured at different angles to cover the required Q-range. This is achieved on reactor sources by varying the sample-to-detector distance, using a moveable detector. On spallation sources, the neutron wavelength varies, and is determined by TOF method, so the position of the detector is fixed. Figures 4.3 and 4.4 show schematic layout of two typical instruments. More technical details can be found elsewhere [2,3,5].

4.1.4 Scattering theory

Scattering events arise from radiation-matter interactions and produce interference patterns that give information about spatial and/or temporal correlations within the sample. Different modes of scattering may be produced: as mentioned before, scattering may be *elastic* or *inelastic*, but also *coherent* or *incoherent*. Coherent scattering from ordered nuclei produces patterns of constructive and destructive interference that contain structural information, while incoherent scattering results from random events and can provide dynamic information. In SANS, only coherent elastic scattering is considered and incoherent scattering, which appears as a background, can be easily measured and subtracted from the total scattering.

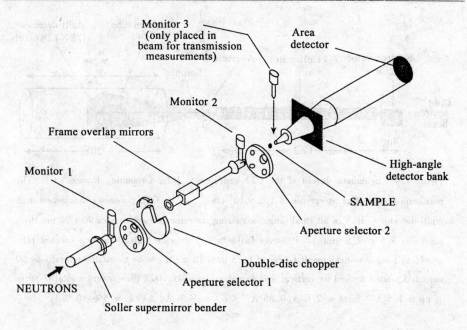

Figure 4.3 Schematic layout of the LOQ instrument, ISIS, Didcot, U.K [2]. After interaction with the sample (typical neutron flux at sample $= 2 \times 10^5 \text{cm}^{-2} \cdot \text{s}^{-1}$), the beam passes into a vacuum tube containing an ^{3}H gas filled detector (active area 64×64 cm^2 with pixel size 6×6 mm^2) placed 4.5 m from the sample. Incident wavelengths range ~ 2.2-10 Å, and the scattering angle $<7°$ gives a useful Q-range of 0.009-0.249 Å^{-1}.

Neutrons interact with the atomic nucleus via strong nuclear forces operating at very short range ($\sim 10^{-15}$ m), i.e., much smaller than the incident neutron wavelength ($\sim 10^{-10}$ m). Therefore, each nucleus acts as a point scatterer to the incident neutron beam, which may be considered as a plane wave. The strength of interaction of free neutrons with the bound nucleus can be quantified by the *scattering length*, *b*, of the atom, which is isotope dependent. In practice, the mean coherent neutron *scattering length density*, ρ_{coh}, abbreviated as ρ, is a more appropriate parameter to quantify the scattering efficiency of different components in a system. ρ represents the scattering length per unit volume of substance and is the sum over all atomic contributions in the molecular volume V_m:

$$\rho_{coh} = \frac{1}{V_m} \sum_i b_{i,coh} = \frac{DN_a}{M_w} \sum_i b_{i,coh} \qquad (4.1.3)$$

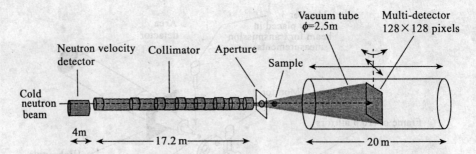

Figure 4.4 Schematic layout of the D22 instrument, ILL, Grenoble, France [1]. The maximum neutron flux at sample is 1.2×10^8 cm$^{-2}\cdot$s^{-1}. D22 possesses the largest area multi-detector (^{3}He) of all small-angle scattering instruments (active area 96×96 cm^2 with pixel size 7.5×7.5 mm^2). It moves inside a 2.5 m wide and 20 m long vacuum tube providing sample-to-detector distances of 1.35 m to 18 m; it can be translated laterally by 50 cm, and rotated around its vertical axis to reduce parallax. D22 thus covers a total Q-range of up to 1.5 Å^{-1} for $\lambda = 2.6$ Å (0.85 Å^{-1} for $\lambda = 4.6$ Å, $\Delta\lambda/\lambda = 5\%$-10%).

where $b_{i,coh}$ is the coherent scattering length of the ith atom in the molecule of mass density D, and molecular weight M_w. N_a is Avogadro's constant. Some useful scattering lengths are given in Table 4.1, and scattering length density for selected molecules in Table 4.1 [6]. The difference in b values for hydrogen and deuterium is significant, and this is exploited in the contrast-variation technique to allow different regions of molecular assemblies to be examined; i.e., one can "see" proton-containing hydrocarbon-type material dissolved in heavy water D_2O.

Table 4.1   Selected values of coherent scattering length, b [6]

Nucleus	$b / (10^{-12}$ cm$)$
^{1}H	$-0.374,1$
^{2}H (D)	$0.667,1$
^{12}C	$0.664,6$
^{16}O	$0.580,3$
^{19}F	$0.565,0$
^{23}Na	$0.358,0$
^{31}P	$0.513,1$
^{32}S	$0.284,7$
Cl	$0.957,7$

Table 4.2 Coherent scattering length density of selected molecules, ρ, at 25℃ [6]. ^{a} Value calculated for the deuterated form of the surfactant ion only (i.e., without sodium counterions), and where the tails only are deuterated

Molecule		$\rho / (10^{10}\ cm^{-2})$
Water	H_2O	−0.560
	D_2O	6.356
Heptane	C_7H_{16}	−0.548
	C_7D_{16}	6.301
AOT	$(C_8H_{17}COO)CH_2CHSO_3^-$ (Na^+)	0.542
	$(C_8D_{17}COO)CH_2CHSO_3^-$ (Na^+)	5.180^a

In neutron scattering experiments, the intensity of the scattered wave, I is measured as a function of a scattering angle, θ, which in the case of SANS is usually less than 10°. Figure 4.5 illustrates schematically a SANS experiment. The incident wave is a plane wave, whose amplitude can be written as [7]

$$A_{in} = A_o \cos(\underline{k}_o \cdot \underline{R} - \Omega_o t) \qquad (4.1.4)$$

where A_o is the original amplitude, $\underline{k}_o$ is the wave vector of magnitude $\frac{2\pi}{\lambda}$, $\underline{R}$ is a position vector, Ω_o is the frequency, and t the time. In static experiments, where relative motions of molecules are ignored, there is no time dependence, and if complex amplitudes are considered, equation 4.1.4 reduces to

$$A_{in} = A_o \exp(i\underline{k}_o \cdot \underline{R}) \qquad (4.1.5)$$

When this wave hits an atom, a fraction of it is scattered, radiating spherically around the scattering centre:

$$A_{sc} = \frac{A_o b}{r} \exp(i\underline{k}_o \cdot \underline{R}) \qquad (4.1.6)$$

where b is the scattering length and r the distance between two point scattering nuclei (Figure 4.6(a)). If the atom is not at the origin but at a position vector $\underline{R}$, the wave scattered in the direction of $\underline{k}_s$ will be phase shifted by $Q \cdot R$ with respect to the incident wave (Figure 4.6(b)). Q is the scattering vector and relates to the scattering angle θ via:

$$Q = \underline{k}_s - \underline{k}_o \qquad (4.1.7)$$

and the magnitude of Q is given by the cosine rule:

$$Q^2 = k_o^2 + k_s^2 - 2k_o k_s \cos\theta \qquad (4.1.8)$$

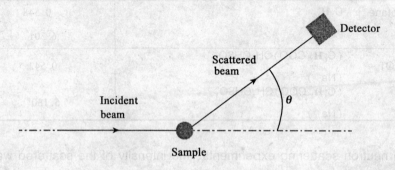

Figure 4.5 Schematic instrumental setup of a small-angle scattering experiment. Sample-to-detector distance is usually 1-20 m; scattering angle $\theta < 10°$.

For coherent elastic scattering, $|k_o| = |k_s| = \dfrac{2\pi n}{\lambda}$, where n is the refractive index of the medium, which for neutron is ~ 1, so $|Q|$ can be obtained by simple geometry as

$$|Q| = Q = 2|k_o| \sin\frac{\theta}{2} = \frac{4\pi}{\lambda}\sin\frac{\theta}{2} \qquad (4.1.9)$$

The magnitude Q has dimensions of reciprocal length and units are commonly Å^{-1}. Large structures scatter to low Q (and angle) and small structures at higher Q values.

Accordingly, the amplitude of the scattered wave at angle θ for an atom at position R from the origin is

$$A_{sc} = \frac{A_o b}{r}\exp[i(\underline{k}_o r - Q \cdot R)] \qquad (4.1.10)$$

Equation 4.1.10 is only valid for the simple case where two point scatterers are considered. In the more realistic case of a very large ensemble of atoms present, the total scattered amplitude is then written as

(a)

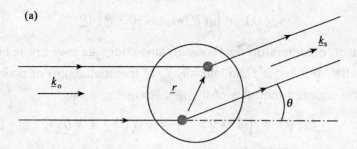

(b)

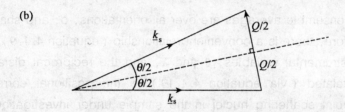

Figure 4.6 Geometrical relationships in scattering experiments. (a) Phase difference between two point scatterers spatially related by the position vector r. The incident and scattered radiation have wave vector k_o and k_s, respectively. For elastic scattering $|k_o| = |k_s|$ $= 2\pi n/\lambda$. (b) Determination of the scattering vector $Q = k_s - k_o$, of amplitude $Q = (4\pi/\lambda)\sin(\theta/2)$.

$$A_{sc} = \frac{A_o}{r}\exp(i\underline{k}_o r) \sum_i b_i \exp(-i\underline{Q} \cdot \underline{R}_i)] \qquad (4.1.11)$$

In the specific case of SANS and Q-range (distances ~10 to 1,000 Å, scattering vectors Q ~ 0.006 to 0.6 Å^{-1}), dilute samples can be treated as discrete particles dispersed in a continuous medium, and the scattering is controlled by the scattering length density, ρ:

$$\rho(\underline{R}) = \frac{1}{v}\sum_j b_i \delta(\underline{R} - \underline{R}_j) \qquad (4.1.12)$$

where the sum extends over a volume V which is large compared with interatomic distances but small compared to the resolution of the experiment.

Then the scattered amplitude is the Fourier transform of this density in the irradiated volume V:

$$A_{sc}(Q) = \int_v \rho(\underline{R}) \exp(-i\underline{Q} \cdot \underline{R}) d\underline{R} \qquad (4.1.13)$$

Radiation detectors do not measure amplitudes as they are not sensitive to phase shift, but instead the intensity I_{sc} of the scattering (or power flux), which is the squared modulus of the amplitude:

$$I_{sc}(Q) = \left(|A(Q)|^2\right) = \langle A(Q) \cdot A^*(Q) \rangle \qquad (4.1.14)$$

For an ensemble of n_p identical particles, equation 4.1.14 becomes [8]

$$I_{sc}(Q) = n_p \left(\left(\left|A_{sc}(Q)^2\right|\right)_o\right)_s \qquad (4.1.15)$$

where the ensemble averages are over all orientations, o, and shapes, s.

Therefore, there is a convenient relationship (equation 4.1.9) between the two instrumental variables, θ and λ, and the reciprocal distance, Q, which is related (via equation 4.1.13) to the positional correlations r between point scattering nuclei in the sample under investigation. These parameters are related to the scattering intensity $I(Q)$ (equation 4.1.15) which is the measured parameter in a SANS experiment, and contains information on intra-particle and inter-particle structure.

4.2 NEUTRON SCATTERING BY MICELLAR AGGREGATES

For monodisperse homogeneous spherical particles of radius R, volume V_p, number density n_p (cm^{-3}) and coherent scattering length density ρ_p, dispersed in a medium of density ρ_m, the normalised SANS intensity $I(Q)$ (cm^{-1}) may be written as [9]

$$I(Q) = n_p \Delta\rho^2 V_p^2 P(Q,R) S(Q) \qquad (4.2.1)$$

where $\Delta\rho = \rho_p - \rho_m$ (cm^{-2}). The first three terms in equation 4.2.1 are independent of Q and account for the absolute intensity of scattering. A so-called *scale factor*, S_F, can then be defined where:

$$S_F = n_p(\rho_p - \rho_m)^2 V_p^2 = \phi_p \cdot \Delta\rho^2 \cdot V_p \qquad (4.2.2)$$

where ϕ_p is the volume fraction of particles. The scale factor is a measure of the validity and consistency of a model used when analysing SANS data; i.e., the S_F value obtained from model fitting can be compared to the expected value, based on sample composition (from equation 4.2.2). The last two terms in equation 4.2.1 are Q-dependent functions. $P(Q,R)$ is the single particle form factor arising from intra-particle scattering. It describes the angular distribution of the scattering due to the particle shape and size. $S(Q)$ is the structure factor arising from inter-particle interactions. To better understand the influence of each term, two scattering profiles are illustrated on Figure 4.7 for the cases of repulsive and attractive forces between interacting homogeneous spheres [8]. It shows how $P(Q)$ and $S(Q)$ can combine to give the overall intensity $I(Q)$. These scattering functions are briefly discussed below.

4.2.1 Single particle form factor $P(Q)$

$P(Q)$ is the function from which information on the size and shape of particles can be obtained. An approximate representation of the form factor $P(Q,R)$ for spheres is shown in Figure 4.7. In general, it appears as a decay although under high resolution maxima and minima are expected at high Q values. The function $P(Q)$ is usually defined as 1.0 at $Q = 0$. General expressions of $P(Q)$ are known for a wide range of different shapes such as homogeneous spheres, spherical shells, cylinders, concentric cylinders and discs [7]. For a sphere of radius R:

$$P(Q,R) = \left[\frac{3(\sin QR - QR\cos QR)}{(QR)^3}\right]^2 \qquad (4.2.3)$$

For certain systems such as microemulsions, a polydispersity function may be introduced to account for the particle-size distribution. For spherical droplets, this contribution may be represented by a Schultz distribution function $X(R_i)$ [10, 11]. Defined by an average radius R^{av} and a root mean square deviation $\sigma = \frac{R^{av}}{(Z+1)^{1/2}}$ with Z a width parameter. $P(Q,R)$ may then be expressed as

$$P(Q,R) = \left[\sum_i P(Q,R_i)X(R_i)\right] \qquad (4.2.4)$$

4.2.2 Structure factor $S(Q)$

The inter-particle structure factor $S(Q)$ depends on the type of interactions in the system, i.e., attractive, repulsive or excluded volume. For spherical particles with low attractive interactions, a reasonable first

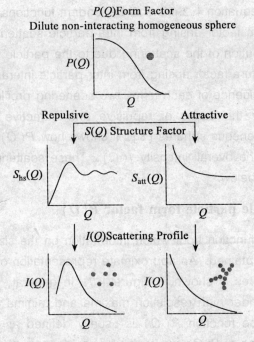

Figure 4.7 Schematic representation of the particle form $P(Q,R)$ and structure $S(Q)$ factors for attractive and repulsive homogeneous spheres, and their contribution to the scattered intensity $I(Q)$. After [8].

approximation is a hard-sphere potential, $S_{hs}(Q)$, given by [12]

$$S_{hs}(Q) = \frac{1}{1 - n_p \cdot f(R_{hs}\phi_{hs})} \quad (4.2.5)$$

where $R_{hs} = R_{core}^{av} + t$ is the hard-sphere radius (with t the hydrocarbon layer thickness) and $\phi_{hs} = \frac{4}{3}\pi R_{hs}^3 n_p$ is the hard-sphere volume fraction. The intensity of scattering, equation 4.2.1 can then be rewritten as

$$I(Q) = \phi_p \Delta\rho^2 V_p [\sum_i P(Q,R_i) X(R_i)] S(Q, R_{hs}, \phi_{hs}) \quad (4.2.6)$$

As shown in Figure 4.7, $S_{hs}(Q)$ is important at low Q values where it reduces the scattering intensity and produces a peak in $I(Q)$ profile at $Q_{max} = 2\pi/D$, with D the mean nearest neighbour distance in the sample. For dilute, non-interacting, systems $\phi_{hs} \prod 0$, so the structure factor disappears, i.e., $S(Q) \prod 1$. For interacting systems, an effective way of reducing $S(Q)$ is by diluting the system [13], or for charged particles by adding salt

[14].

For systems where attractive interactions have to be considered, particularly in the vicinity of phase separation regions or cloud point in binary phase diagrams, a structure factor known as the Ornstein-Zernike (OZ) expression may be used [8]:

$$S_{OZ}(Q) = 1 + \frac{S(0)}{1 + (Q\xi)^2} \qquad (4.2.7)$$

where $S(0) = n_p k_B T_\chi$, with k_B Boltzmann's constant, T the temperature, and χ the isothermal compressibility. ξ is a correlation length. Far from phase boundaries, $S(0) \Pi 0$, and so $S_{OZ}(Q)$ disappears, i.e., $S(Q) \Pi 1$.

4.2.3 Neutron contrast variation

As mentioned previously, the very different neutron scattering lengths of hydrogen and deuterium are exploited in SANS experiments to reveal details of structure and composition at interfaces. This is routinely applied in microemulsion droplets where different regions can be highlighted by selectively varying the scattering length density of the surfactant, oil or aqueous phase. Three contrasts are commonly studied—core, shell and drop —which can be fitted individually or simultaneously [15]. Figure 4.8 illustrates the scattering length density profiles for a water-AOT-n-heptane microemulsion for the three contrasts. The initial situation, where all components are hydrogenated, is shown in Figure 4.8(a). As reported in Table 4.2, the scattering length densities of H_2O, n-heptane and AOT are very similar, so that deuteration of the water and/or oil phases allows contrast match of specific regions within the system. The distance from the droplet centre is Z, and so ρ depends on Z owing to the presence of the different materials. Apart from a few subtle effects, such as hydrogen bonding, this isotopic exchange does not usually affect the chemical or physical properties of the system significantly.

4.2.4 SANS approximations

A first estimation of the size and shape of particles can be obtained from simple relations between $I(Q)$ and the particle radius (or thickness) based

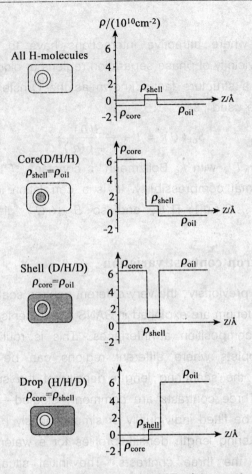

Figure 4.8 Elucidation of the structure of water-AOT-n-heptane microemulsion droplets by contrast variation. The scattering length density, ρ, depends on Z, the distance from the centre of a droplet.

on a few assumptions and/or approximations.

Guinier approximation

The SANS profile $I(Q)$ is very sensitive to different particle shapes. In particular, the Guinier approximation relates the low Q part of the scattering plot to a radius of gyration R_g of the particle. At low Q (Guinier regime), the single particle form factor $P(Q,R)$ for dilute systems simplifies to [16]

$$P(Q,R) = 1 - \frac{Q^2 R_g^2}{3} \qquad (4.2.8)$$

where R_g is the root mean square value of the radius averaged over the volume of particle, and relates to the shape of the particle:
- For spheres or cylinders

$$R_g = \left(\frac{3}{5}\right)^{1/2} R \qquad (4.2.9)$$

- For thin discs

$$R_g = \frac{R}{4^{1/2}} \qquad (4.2.10)$$

- For long rods

$$R_g = \frac{L}{12^{1/2}} \qquad (4.2.11)$$

Where, R is the radius of the spheres or cylinders, or disc thickness, and L is the rod length.

Assuming $S(Q) = 1$ and $1 - X^2 \approx \exp(-X^2)$, Equation 4.2.1 becomes

$$I(Q) \approx \phi_p \Delta\rho^2 V_p \exp\left(-\frac{Q^2 R_g^2}{3}\right) \qquad (4.2.12)$$

The Guinier plot—i.e., $\ln I(Q)$ versus Q^2—should includes a linear section up to the limit $QR_g < 1$. The associated slope is $-\frac{R_g^2}{3}$, and so R_g can be determined for any isometric particles. Another useful expression for the Guinier approximation is [7, 17]

$$I(Q) \propto Q^{-D} \exp\left(-\frac{Q^2 R^2}{K}\right) \qquad (4.2.13)$$

Equations 4.1.11 and 4.1.12 are equivalent. They are valid for non-interacting particles (i.e., $S(Q) \prod 1$) only, and over a restricted Q-range. The proportionality constant depends on the concentration and isotopic composition. The exponent D is 1 for cylinders, 2 for discs, and 0 for spheres. R is the characteristic dimension of the particle, i.e., the cross sectional radius for cylinders, the thickness for discs, and the radius for spheres. K is an integer of value 4 for cylinders, 12 for discs, and 5 for spheres. Depending on the geometry, the dimension R can be obtained by plotting different quantities against Q^2:

- $\ln[I(Q) \cdot Q]$ vs. Q^2:   cylinder radius $= \sqrt{\text{slope} \times 4}$ (4.2.14)
- $\ln[I(Q) \cdot Q^2]$ vs. Q^2:   disk thickness $= \sqrt{\text{slope} \times 2}$ (4.2.15)

- $\ln[I(Q)]$ vs. Q^2 ($QR < 1$):   sphere radius $= \sqrt{\text{slope} \times 5}$ (4.2.16)

Therefore the most probable particle shape can be predicted by comparison of the three different $I(Q) \cdot Q^D$ vs. Q^2 plots (i.e., the one giving a linear decay).

4.2.5 Porod approximation

At high Q values, the SANS intensity is sensitive to scattering from local interface rather than the overall inter-particle correlations. Then $I(Q)$ is related to the total interfacial area S, and the asymptotic intensity (see Figure 4.9) may be analysed using the Porod equation [18, 19]:

$$I(Q) = 2\pi\Delta\rho^2 \left(\frac{S}{V}\right) Q^{-4} \qquad (4.2.17)$$

where S/V is the total interfacial area per unit volume of solution (cm^{-1}). The Porod equation is only valid for smooth interfaces and a Q-range $\gg 1/R$ (Porod regime). Assuming all the surfactant molecules are located at the interface, the average area per surfactant head group, a_s, can be estimated from

$$a_s = \left(\frac{S/V}{N_s}\right) \qquad (4.2.18)$$

where N_s is the number density of surfactant molecules (i.e., surfactant concentration $\times$ Avogadro's number). The Porod approximation can also be used to estimate the particle radius [8]. For monodisperse spheres of radius R, a plot of $[I(Q) \cdot Q^4]$ vs. Q gives a first maximum at $Q \approx 2.7/R$ and a minimum at $Q \approx 4.5/R$ (see Figure 4.9).

The Guinier and Porod approximations thus offer simple relations that allow a first estimation of the size and shape of colloidal particles. However, they are limited to dilute, non-interacting systems. As mentioned in the previous section, dilution or addition of salt allow the screening of interactions, so that the assumption $S(Q) = 1$ in the low Q-range becomes valid and the Guinier approximation can be applied. For microemulsions, these conditions do not always hold, they might be unstable to dilution, and also addition of salt may introduce structural changes. In such cases, information about the size and shape of aggregates are obtained by fitting SANS experimental data to more complex mathematical models, such as those derived for polydisperse spherical droplets and introduced in this section. More details about these models can be found elsewhere [9, 20].

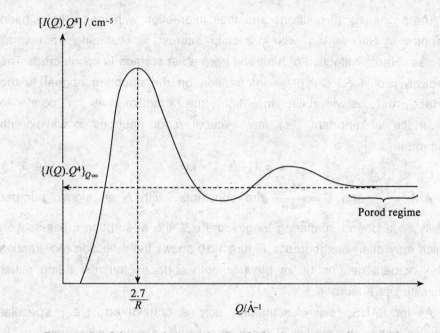

Figure 4.9 Schematic diagram of a Porod plot for near-monodisperse spheres (see text for details).

4.3 NEUTRON REFLECTION

Neutron reflection (NR) is a very useful and reliable method since it provides a direct measure of the surface excess, and also permits structural features of the interface to be elucidated. Just like SANS the only disadvantage is the necessity to carry out expensive experiments at neutron facilities, and in some cases the need for deuterated solvents and/or surfactants. Tensiometry, on the other hand, is a very accessible method but only provides indirect determination of the surface excess via surface tension measurements and application of the Gibbs equation (see Section 2.1.2). Certain material and basic definitions relevant to NR have already been covered in relation to small-angle scattering, and where necessary reference is made.

4.3.1 Background theory

The properties of neutrons and their interaction with matter have been presented in Section 4.1 with particular interest in scattering from small particles. Here, reflection of neutrons from a flat surface is considered. The reflectivity profile $R(Q)$ gives information on the structure normal to the interface, and, as with reflection of light, the refractive index, n, normal to the surface is important. For any material n for neutrons is wavelength dependent [21], i.e.,

$$n = 1 - \lambda^2 A + i\lambda C \qquad (4.3.1)$$

where $A = \dfrac{Nb}{2\pi}$ and $C = \dfrac{N\sigma_{abs}}{4\pi}$ are constants, with N an atomic number density, b a bound scattering length and σ_{abs} the absorption cross-section (which may often be ignored). Figure 4.10 shows the reflection of a fraction of the incident neutron beam by a smooth surface; the rest being either transmitted or adsorbed.

As for SANS, elastic scattering only is considered, i.e., specular reflection, when the moduli of the incident and reflected wave vectors, k_o and k respectively, are equivalent ($|k_o| = |k|$). A scattering vector Q_z defined in one dimension only, i.e., the z direction perpendicular to the sample surface, is then given as

$$Q_z = \frac{4\pi n}{\lambda}\sin\theta \qquad (4.3.2)$$

where n is the refractive index. The reflected intensity $R(Q_z)$ is thus measured as a function of Q_z either by varying the wavelength of the neutron beam, λ, and keeping the angle of incidence, θ, constant (method at pulsed neutron sources), or by selecting a constant λ value and varying θ (at reactor sources).

For a plane wave incident upon a surface, if the first medium is air, $n_0 \approx 1$, total external reflection occurs below a critical incidence angle $\theta_0 = \theta_c$, and a critical value Q_c, defined by the wavelength λ_c or angle θ_c, is reached. For a clean D_2O surface (Figure 4.11(a)) if $\theta < \theta_c$ (i.e., below Q_c) then there is total reflection and $R(Q) \equiv 1$, whilst above Q_c the reflectivity falls off

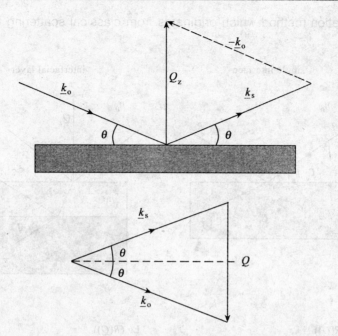

Figure 4.10 Geometry of a neutron reflection experiment and direction of the scattering vector Q, where k_o and k_s are the incident and scattered wave vectors, and θ is the scattering angle.

sharply as Q^{-4}. The region where $R(Q) = 1$ is used to determine the instrument calibration scale factor. In the case of a surfactant monolayer on a water sub-phase (Figure 4.11(b)), Q_c is usually reached when measurements are made at $\theta < 1.5°$ [22]. On passing through an adsorbed layer the incident neutron beam is partially transmitted and reflected. Waves are reflected from both top and bottom surfaces of the thin interfacial film. There is then an interference between these two reflected beams, resulting in the appearance of a "fringe" in the $R(Q)$ profile. The position of a minimum Q_{min} is related to the layer thickness τ by $Q_{min} \approx 2\pi/\tau$ (Figure 4.11(b)).

4.3.2 Layer thickness and adsorbed density

The analysis of specular reflectivity in single- or multiple-layer systems can be done either by comparison with a reflectivity profile calculated using an exact optical matrix method, or by using the kinematic (or Born) approximation method which originates from classical scattering theory.

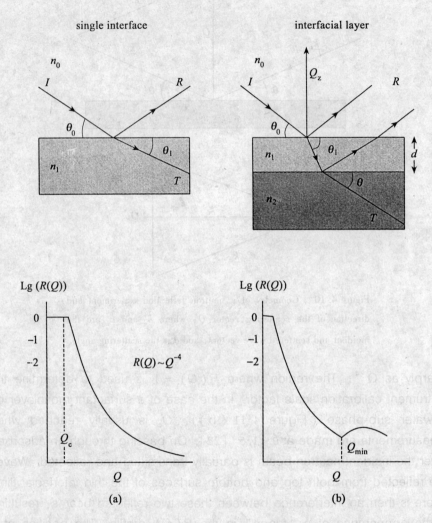

Figure 4.11 Specular reflection at the sharp interface between two bulk media (a), and for a thin interfacial layer (b) of thickness d sandwiched between two media. I, R and T are the incident, reflected and transmitted beams respectively. Other symbols as defined in the text. Also illustrated are schematic $R(Q)$ profiles for the two situations.

In the first approach, a characteristic matrix per layer is defined that relates electric vectors in successive layers in terms of Fresnel reflection coefficients— combining refractive index and reflected angle —and phase factors introduced on traversing each layer [23]. A detailed account of this approach can be found in several texts [e.g. 21, 24]. The resulting model typically consists of a series of layers, each with a scattering length density ρ and thickness τ, into which an interfacial roughness between any two consecutive layers can also be incorporated. Indeed most surfaces are affected by a local roughness that reduces the specular reflectivity [25, 26], especially for liquid surfaces where thermally excited capillary waves are present. The calculated and measured profiles are compared, and ρ and τ for each layer varied until the optimum fit is found (determined by a least-squares iterative fitting process). Then secondary parameters such as the area per molecule or the coverage can be easily determined (see below).

Consider the simple situation of a single uniform layer, of thickness d and refractive index n_1, introduced between bulk media of refractive index n_0 and n_2 (Figure 4.11(b)). If the surfactant solution is made up in null reflecting water (NRW), a mixture of 8 mol % D_2O in H_2O where $\rho = 0$ and hence $n = 1$. Then there is no scattering contribution from the NRW, and reflection arises purely from the surfactant monolayer. Fitting a single-layer model to the data immediately allows the surface excess Γ to be calculated through [27, 28]

$$A_s = \frac{\sum b_i}{\rho(z)\tau} = \frac{1}{\Gamma N_a} \qquad (4.3.3)$$

where $\sum b_i$ is the sum of scattering lengths over a single molecule, N_a is Avogadro's number, and $\rho(z)$ and τ are the optimised scattering length density and layer thickness determined by fitting. The absolute value of τ is model dependent and will vary with the choice of the distribution function employed to describe the scattering length density profile normal to the surface. However, as described by Simister et al. [29], the sets of $\rho(z)$ values required to give a good fit exactly compensate for the change in τ so that A_s is independent of the uncertainty in τ.

A second analysis method is the kinematic (or Born) approximation [30]. In this approach, the reflectivity $R(Q_z)$ is related to the scattering length density profile normal to the interface, $\rho(z)$, by

$$R(Q_z) = \frac{16\pi^2}{Q_z^2}|\hat{\rho}(Q_z)|^2 \qquad (4.3.4)$$

where $\hat{\rho}(Q_z)$ is the one-dimensional Fourier transform of $\rho(z)$:

$$\hat{\rho}(Q_z) = \int_{-\infty}^{+\infty} \exp(-iQ_z z)\rho(z)\,dz \qquad (4.3.5)$$

4.3.3 Partial structure factor analysis

Details of the development of the kinematic approximation for the study of adsorbed thin films are described elsewhere [30,31]. For surfactant monolayers adsorbed at the air-solution interface, the main features of interest—in addition to film thickness and surface coverage—are the relative positions of the chains, head, and water, and the widths of their distributions normal to the interface. Such structural features can be obtained through analysis of complementary $R(Q)$ profiles, determined at different isotopic compositions using hydrogen/deuterium labelling of the surfactant. This is known as a partial structure factor (PSF) analysis, which is valid under the kinematic approximation. Depending on the labelling scheme, the total scattering may be expressed in terms of numerous partial structure factors that are descriptive of the various components in the interface [30, 32]. Below is a brief description of such analysis where a simple binary system consisting of a surfactant solute, A, and a solvent, S, are considered. Then the scattering length density may be represented as

$$\rho(z) = N_S(z)b_S + N_A(z)b_A \qquad (4.3.6)$$

where $N_A(z)$ and $N_S(z)$ are the number densities of solute and solvent respectively and b_i are the scattering lengths. Combining equations 4.3.4 and 4.3.6 gives

$$R(Q_z) = \frac{16\pi^2}{Q_z^2}[b_A^2 h_{AA} + b_S^2 h_{SS} + 2b_A b_S h_{AS}] \qquad (4.3.8)$$

where the h_{ii} and h_{ij} are the PSFs : h_{ii} are self-terms that contain information about distributions of the individual components, and are one-dimensional Fourier transforms of $N_i(z)$:

$$h_{ii}(Q_z) = |N_i(Q_z)|^2 \qquad (4.3.9)$$

h_{ij} are cross-terms that describe the relative positions of the different components. In the example given here, of a surfactant solute and solvent, assuming that the distributions of A and S at the interface are exactly even (symmetrical about the centre) and odd respectively [33], then the following relationship holds,

$$h_{AS} = \pm (h_{AA} h_{SS})^{1/2} \sin(Q_z \delta_{AS}) \qquad (4.43)$$

where δ is the separation between the centres of the surfactant and solvent distributions. The distributions may not be exactly even/odd, and deviations from this assumption may affect the accuracy with which δ_{AS} can be determined. Circumstances where this approximation fails has been discussed in full elsewhere [34], but this is not expected to arise for simple surfactants [35, 36].

In principle, the $N_i(z)$ can be obtained by Fourier transformation of the PSFs, but in practice, an analytical function that best represents the form of $N_i(z)$ is assumed. The function is then Fourier transformed and fitted to the experimental data. For monolayer of soluble surfactants, it is shown that a Gaussian distribution is a good representation of the number density profile, $N_A(z)$, [37]:

$$N_A(z) = N_{A0} \exp\left(\frac{-4z^2}{\sigma_A^2}\right) \qquad (4.3.10)$$

where N_{A0} is the maximum number density and σ_A is the full width at the $1/e$ of the maximum number density. The total adsorbed amount in the monolayer, Γ_m, is related to N_{A0} through

$$\Gamma_m = \frac{1}{A_s} = \frac{\sigma_A N_{A0} \pi^{1/2}}{2} \qquad (4.3.11)$$

where A_s is the area per molecule.

For the interfacial solvent distribution, a convenient analytical form is the tanh function given by

$$N_S = N_{S0}\left[\frac{1}{2} + \frac{1}{2}\tanh\left(\frac{z}{\xi}\right)\right] \qquad (4.3.12)$$

where N_{S0} is the number density of water in the bulk solution and ξ is the width parameter. The respective PSFs for distributions described by equations 4.3.10 and 4.3.12 are then

$$Q_z^2 h_{AA} = \frac{\pi \sigma_A^2 N_{A0}^2 Q_z^2}{4} \exp\left(-\frac{Q_z^2 \sigma_A^2}{8}\right) \qquad (4.3.13)$$

$$Q_z^2 h_{SS} = \frac{N_{S0}^2 \zeta^2 \pi^2 Q_z^2}{4} \cos\text{ech}^2\left(\frac{\zeta \pi Q_z}{2}\right) \quad (4.3.14)$$

The cross PSF, h_{AS}, can be obtained directly by combining the above equations for h_{AA} and h_{SS} into equation 4.3.7 and fitting the resulting function for δ_{AS}:

$$Q_z^2 h_{AS} = \frac{\sigma_A N_{S0} N_{A0} \pi^{3/2} \zeta Q_z^2}{4} \exp\left(-\frac{Q_z^2 \sigma_A^2}{16}\right) \cos\text{ech}\left(\frac{\zeta \pi Q_z}{2}\right) \sin\pi\delta_{AS}$$

$$(4.3.15)$$

Structural parameters (with the exception of δ_{AS}) obtained from the kinematic approximation are dependent upon the assumed distribution shapes. For the solute, however, it can be shown that the surface coverage, Γ_m, is independent of any assumptions made about the $N_A(z)$ distribution [30]. For a monolayer with a Gaussian distribution at the surface of water that is contrast matched to air, NRW, the reflectivity is given by [38, 39]

$$\frac{Q_z^2 R(Q_z)}{16\pi^2} \approx (\Gamma_m N_a b_A)^2 \exp(-Q_z^2 \sigma_A^2) \quad (4.3.16)$$

where N_a is the Avogadro constant. Hence, a plot of $\ln Q_z^2 R(Q_z)$ vs. Q_z^2 extrapolated to $Q_z = 0$ yields a model independent value for Γ_m.

The kinematic approximation assumes all scattering is due to single events, i.e., effects of multiple scattering within the sample are ignored. Therefore this approximation is only valid when the scattering is weak, that is, the incident intensity, I_o is much greater than the scattered intensity, I_s. When applied to reflection, the approximation breaks down in the region $Q \sim Q_c$ since $I_s \sim I_o$, and the scattering is no longer "weak". Thus interpretation of reflectivity data using the kinematic theory is strictly only valid far from the region of total reflection. As a result, equation 4.3.4 is only approximate and fails at low Q. To utilize the entire range of reflectivity data available, and to account for the failure of the kinematic approximation as $Q_z \sim Q_c$, each calculated $R(Q_z)$ needs to be corrected before being compared with the observed reflectivity [28, 40].

An excellent review of the applications of neutron reflectometry has been published [41].

REFERENCES

1. Bacon, G. E. '*Neutron Scattering in Chemistry*', Butterworths, 1977, London.
2. ILL World Wide Web page: http://www. ill. fr
3. ISIS World Wide Web page: http://www. isis. rl. ac. uk
4. Finney, J. ; Steigenberger, S. *Physics World* 1997, *10*, issue 12 (report published on PhysicsWorld Web page: http://www. physicsweb. org/article/world/10/12/7).
5. Heenan, R. K; Penfold, J. ; King, S. M. *J. Appl. Cryst.* 1997, *30*, 1140.
6. King, S. M. '*Small-Angle Neutron Scattering*', a report published on ISIS Web page, 1997.
7. Cabane, B. in '*Surfactant Solutions: New Methods of Investigation*', (Ed. Zana, R.), Surfactant Science Series vol 22, p57-139, Marcel Dekker Inc. , 1987, New York.
8. Eastoe, J. in '*New Physico-Chemical Techniques for the Characterisation of Complex Food Systems*', (Ed. Dickinson, E.), p268-294, Blackie, 1995, Glasgow.
9. Ottewill, R. H. in '*Colloidal Dispersions*', (Ed. Goodwin, J. W.), R. S. C, 1982, London.
10. Chen, S-H. *Ann. Rev. Phys. Chem.* 1986, *37*, 351.
11. Kotlarchyk, M. ; Chen, S-H. ; Huang, J. S. ; Kim, M. W. *J. Phys. Chem.* 1984, *29*, 2054.
12. Ashcroft, N. W. ; Lekner, J. *Phys. Rev.* 1966, *145*, 83.
13. Cebula, D. J. ; Myers, D. Y. ; Ottewill, R. H. *Colloid Polym. Sci.* 1982, *260*, 96.
14. Goodwin, J. W. ; Ottewill, R. H. ; Owens, S. M. ; Richardson, R. A. ; Hayter, J. B. *Macromol. Chem. Suppl.* 1985, *10/11*, 499.
15. Heenan, R. K. ; Eastoe, J. *J. Appl. Cryst.* 2000, *33*, 749.
16. Guinier, A. *Annales de Physique* 1939, *12*, 161.
17. Porte, G. in '*Micelles, Membranes, Microemulsions, and Monolayers*' (Eds. Gelbart, W. M. ; Ben-Shaul, A. ; Roux, D.), Springer-Verlag, 1994, New York, p105-145.

18. Porod, G. *Koll. Z.* 1951, *124*, 82.
19. Auvray, L. ; Auroy, P. in '*Neutron, X-Ray and Light-Scattering*' (Eds. Lindner, P. ; Zemb, Th.), Elsevier Science Publishers, 1991, Holland.
20. Markovic, I. ; Ottewill, R. H. ; Cebula, D. J. ; Field, I. ; Marsh, J. *Colloid Polym. Sci.* 1984, *262*, 648.
21. Born, M. ; Wolf, E. , '*Principles of Optics*', 6th edition, Pergamon Press, 1980, Oxford.
22. Eastoe, J. '*Small-Angle Neutron Scattering*' in '*New Physico-Chemical Techniques for the Characterisation of Complex Food System*', (ed. Dickinson, E.), p268-294, Blackie, 1995, Glasgow.
23. Abelès, F. *Ann. Phys. (Paris)* 1948, *3*, 504.
24. Heavens, O. J. '*Optical Properties of Thin Solid Films*', Butterworths Scientific Publications, 1955, London.
25. Nevot, L. ; Croce, P. *Rev. Phys. Appl.* 1980, *15*, 761.
26. Sinha, S. K. ; Sirota, E. B. ; Garoff, S. ; Stanley, H. B. *Phys. Rev. B* 1988, *38*, 2297.
27. Thomas, R. K. ; Penfold, J. *J. Phys. Condens. Matter* 1990, *2*, 1369.
28. Lu, J. R. ; Simister, E. A. ; Lee, E. M. ; Thomas, R. K. ; Rennie, A. R. ; Penfold, J. *Langmuir* 1992, *8*, 1837.
29. Simister, E. A. ; Thomas, R. K. ; Penfold, J. ; Aveyard, R. ; Binks, B. P. ; Cooper, P. ; Fletcher, P. D. I. ; Lu, J. R. ; Sokolowski, A. *J. Phys. Chem.* 1992, *96*, 1383.
30. Crowley, T. ; Lee, E. M. ; Simister, E. A. ; Thomas, R. K. *Physica B* 1991, *173*, 143.
31. Lu, J. R. ; Simister, E. A. ; Thomas, R. K. ; Penfold, J. *J. Phys. Chem.* 1993, *97*, 6024.
32. Simister, E. A. ; Lee, E. M. ; Thomas, R. K. ; Penfold, J. *J. Phys. Chem.* 1992, *96*, 1373.
33. Cooke, D. J. ; Lu, J. R. ; Lee, E. M. ; Thomas, R. K. ; Pitt, A. R. ; Simister, E. A. ; Penfold, J. *J. Phys. Chem.* 1996, *100*, 10298.
34. Simister, E. A. ; Lee, E. M. ; Thomas, R. K. ; Penfold, J.

Macromol. Rep. 1992, *A29*, 155.
35. Li, Z. X. ; Lu, J. R. ; Thomas, R. K. ; Penfold, J. *Prog. Colloid Polym. Sci.* 1995, *98*, 243.
36. Li, Z. X. ; Lu, J. R. ; Thomas, R. K. ; Penfold, J. *J. Phys. Chem. B* 1997, *101*, 1615.
37. Lu, J. R. ; Li, Z. X. ; Smallwood, J. A. ; Thomas, R. K. ; Penfold, J. *J. Phys. Chem.* 1995, *99*, 8233.
38. Li, Z. X. ; Lu, J. R. ; Thomas, R. K. *Langmuir* 1997, *13*, 3681.
39. Li, Z. X. ; Lu, J. R. ; Thomas, R. K. ; Rennie, A. R. ; Penfold, J. *J. Chem. Soc. Faraday Trans.* 1996, *92*, 565.
40. Crowley T. L. *Physica A* 1993, *195*, 354.
41. Penfold, J. ; Richardson, R. M. ; Zarbakhsh, A. ; Webster, J. R. P. ; Bucknall, D. G. ; Rennie, A. R. ; Jones, R. A. L. ; Cosgrove, T. ; Thomas, R. K. ; Higgins, J. S. ; Fletcher, P. D. I. ; Dickinson, E. ; Roser, S. J. ; McLure, I. A. ; Hillman, A. R. ; Richards, R. W. ; Staples, E. J. ; Burgess, A. N. ; Simister, E. A. ; White, J. W. *J. Chem. Soc. Faraday Trans.* 1997, *93*, 3899.